A Global Parliament
Essays and Articles

Noteworthy Praise for
A Global Parliament: Essays and Articles

"At a time when many in civil society question the legitimacy of our international architecture, and globalization brings home the reality that no nation alone — mighty or modest — can hope to solve our interconnected challenges, the global community needs to reconsider its conduct of international affairs. Of fundamental importance is how to involve the will of the global polity. Professors Falk and Strauss offer fresh thinking and creative ideas for doing just that. This book is a must read for those who care about our common future."

Mike Moore, Former Prime Minister of New Zealand and former Director General of the World Trade Organisation

"This collection brings together a decade of energetic advocacy for deepening democracy through a global parliament by Richard Falk and Andrew Strauss. It is brimming with insight, imagination and inspiration, but does not neglect the obstacles to and the tough practicalities of such a venture. This book offers both a timely injection of hope about and a blueprint for the creation of international democratic structures."

Hilary Charlesworth, Professor and Director of the Centre for International Governance and Justice, Regulatory Institutions Network at the Australian National University

"Falk and Strauss, as we would expect of them, challenge what others perceive as 'realism', and in this book offer provocative analysis and proposals for the road forward in international relations and law. An important contribution to contemporary debate."

Dame Rosalyn Higgins, former President of the International Court of Justice

"Richard Falk and Andrew Strauss lead the way in calling for a global parliament — a representative body that would have the scope and authority to address vital global concerns. Because democracy functions exclusively within national borders, no democratic authority is presently capable of dealing with the emergent transborder issues that so shape our global age. This needs to be put right, and Falk and Strauss tell us how it can be."

David Held, Graham Wallas Professor of Political Science, London School of Economics

A Global Parliament: Essays and Articles
By Richard Falk and Andrew Strauss

Published by the Committee for a Democratic U.N., Berlin, 2011.

www.kdun.org

ISBN 978-3-942282-08-6

A Global Parliament
Essays and Articles

by Richard Falk and Andrew Strauss

COMMITTEE FOR A DEMOCRATIC U.N.

Contents

Editor's Preface

This collection of articles and essays by Richard Falk and Andrew Strauss on the creation of a global parliament is being published at a time of upheaval. The unprecedented popular uprisings in the Arab world mark an important milestone. In the words of Richard Falk, the outcome of this "fourth rupture in global governance since the fall of the Berlin Wall in 1989," may determine the future "of self-determination in the entire Arab world, and possibly beyond."[1] A considerable strengthening and revival of democracy in the world seems to be within the realm of possibility.

However, in this time of globalization no society is able to escape the impact of globally integrated economic and financial markets or of climate change. This makes democratic self-determination which is limited strictly to the nation-state difficult, if not impossible. As the author of this book's foreword, former UN Secretary-General Boutros Boutros-Ghali, has pointed out, "democracy within the state will diminish in importance if the process of democratization does not move forward at the international level. Therefore, we need to promote the democratization of globalization, before globalization destroys the foundations of national and international democracy."[2]

Falk and Strauss are among the very few academics in the world who have specifically dealt with the question of a popularly elected global body. What Boutros-Ghali has described as the "missing link of democratization" that is "almost completely neglected,"[3] is exactly the subject of this volume.

The article "Toward Global Parliament" that was published by Falk and Strauss in *Foreign Affairs* in 2001 is thus far the one that is probably most cited on the subject. It is the first article reprinted in this volume but not the first that they published on the subject. This collection does not follow a chronological order. Instead, the texts are grouped according to overarching topics.

The reflections and arguments of Falk and Strauss on a Global Peoples Assembly or, synonymously, on a Global Parliamentary Assembly (GPA), are still topical and an indispensable contribution to the debate on global democracy. Some of the basics of their approach are widely shared by advocates of a UN Parliamentary Assembly (UNPA). For instance these include reference to the European Parliament as a model and the strategy of following an incremental approach that starts with a largely consultative assembly and proceeds to the eventual goal of a democratic global legislative system.[4]

But there are differences too. Falk and Strauss for example argue that a small number of twenty to thirty countries that are geographically, culturally and economically diverse could initiate the parliamentary project by agreeing on an intergovernmental treaty that provides for the establishment of a directly elected GPA. Over time, as more and more countries join the project, the assembly would gain more and more global democratic legitimacy.

By contrast, advocates of a UNPA usually envisage that the body could be created by a decision of the UN General Assembly pursuant to Article 22 of the UN Charter, that all UN member states would participate, and that initially it would be composed of delegates chosen from within national and regional parliaments. Direct elections would then be introduced in later stages.

These and other different viewpoints enrich the debate, and it is far from clear which proposals will eventually win the day. With this publication we want to make the work of Falk and Strauss better accessible and hope to generate new interest.

We would like to express our gratitude to all those who have helped to make this publication possible. In particular, we would like to thank Marianne Obermüller and the Earthrise Society in Munich, Germany, for their financial support.

Andreas Bummel
Committee for a Democratic U.N.

---

[1] Richard Falk. "Egypt's Berlin Wall moment." Al Jazeera Online, February 8, 2011. http://english.aljazeera.net/indepth/opinion/2011/02/20112795229925377.html.

[2] Boutros Boutros-Ghali, "Message to the Campaign for the Establishment of a UN Parliamentary Assembly," May 16, 2007. http://www.unpacampaign.org/documents/en/-BBG200705.pdf.

[3] Boutros Boutros-Ghali, "The missing link of democratization." *OpenDemocracy*, June 9, 2009. http://www.opendemocracy.net/article/boutros-boutros-ghali/UN-parliament-global-democracy.

[4] See, for example, Dieter Heinrich, The Case for a United Nations Parliamentary Assembly. Berlin: Committee for a Democratic UN, 2010 [1992]. http://www.kdun.org/88/.

Foreword by
Boutros Boutros-Ghali

During my term as Secretary-General of the United Nations I spent a great proportion of my time dealing with the dual concerns of peace and development. I laid out many of my thoughts and initiatives regarding peace in 1992, when at the request of the Security Council I submitted *An Agenda for Peace*. Similarly, a year later at the request of the General Assembly I laid out my primary thoughts and initiatives regarding development in *An Agenda for Development*.

Trying to lead the way toward the reforms I suggested was quite difficult, and I came to recognize that the key to overcoming reform gridlock in both areas is democracy. Regarding peace, if no democratic institutions exist to channel popular discontent, confrontation and oftentimes violent conflict will result. Likewise, regarding development, all over the world entrenched elites whose power is unchecked by democratic institutions bleed economies dry, and in the absence of institutions of popular accountability, their ability to impede development cannot be effectively countered.

Having recognized the connection between development and peace on the one hand, and democracy on the other, I decided that I should also explore the role of the United Nations in democratization in more detail. It was a bumpy road as my mandate to produce a third agenda was challenged, but finally, two weeks before my departure from the United Nations, I finished my *Agenda for Democratization* and arranged for its distribution in the UN system.

What I argued for in this agenda, among other things, was democratization among nations. "If democratization is the most reliable way to legitimize and improve national governance," I noted, "it is also the most reliable way to legitimize and improve international organizations."[1] How can the United Nations as the world's most universal political organization be a credible and effective promoter of democracy at the national level if it does not pursue the same principle internationally in its own sphere?

Today, fifteen years after the publication of the *Agenda for Democratization,* and in light of the mass movements for democracy, most recently in the Arab world, this question takes on great salience. As more and more decisions with far reaching domestic implications are made at the intergovernmental level, democracy to be effective must extend beyond state borders. How exactly this can be done calls for a great deal of creative thought, but such thought has not been forthcoming. In fact, most commentators on global governance seem hopelessly mired in the existing system. This is why the arrival of this book that gathers the essays and articles by Professors Richard Falk and Andrew Strauss on a Global Parliamentary Assembly is a breath of fresh air.

What is presented in the pages that follow is not only a thoughtful and comprehensive exploration of how a global parliament would help correct for the dysfunction of the present global system, but it is also a clear-headed pragmatic blueprint for achieving such a parliament.

I cannot overstate the importance of this rare inquiry into this most crucial global governance challenge of our time: how to adapt parliamentarianism to the global system. As I argued in the *Agenda for Democratization*, "by carrying the views and concerns of their constituents to the international arena, parliamentarians offer a direct channel for increasing the legitimacy, responsiveness and effectiveness of international organizations."[2]

Over the past fifteen years, the call for increased participation of parliamentarians in international affairs has gathered more and more momentum. In 2007, this culminated in the Campaign for a UN Parliamentary Assembly which I have supported wholeheartedly from the very beginning.

I am convinced that in the years to come the popular struggle for democracy will become a major force on the global stage. It is high time that the establishment of a Global Parliamentary Assembly is put onto the international agenda. When it is, this work by Professors Falk and Strauss will have laid the groundwork for implementation.

---

[1] Boutros Boutros-Ghali, Supplement to reports on democratization. Report to the 51st Session of the United Nations General Assembly. 20 December 1996, para. 66.

[2] Ibid., para. 87.

Introduction

In Retrospect

We wrote the first of our works included in this book as *Op Eds* for the *Philadelphia Inquirer* and the *International Herald Tribune* in 1997. That year of our beginning marked the cresting of an era that had begun in the concluding days of the Cold War and that was uniquely both democratic and internationalist.

Its democratic character was made manifest by widespread regime change. From Eastern Europe to Asia and Africa authoritarian governments yielded to new democratic ones. Though many of the democracies born of this era were deeply flawed and some have not stood the test of time, they commonly represented the hope of the day that history was on the side of greater freedom.

The era's internationalist character was made manifest as well by a boom in the building of international organizations. Though the balance of the boom's affects on social and economic justice as well as the environment has been mixed, it too represented the spirit of the times, and in doing so transformed the global institutional skyline. To name but a few of the period's significant institutional developments, the Maastricht Treaty of 1993 gave birth to the European Union. The North American Free Trade Agreement came into force in 1994. The World Trade Organization was established in 1995, and the Statute of the International Criminal Court was adopted in Rome in the summer of 1998.

These two great post-Cold War trends of democratization and internationalism came together in the emergence of what has come to be called the global democracy movement. Citizen organizations, no longer content to limit their political participation to the domestic arena, came of age in the 1990's as a global political force. Many human rights and environmental organizations, in particular, came to play a role that was different from, but at least equal in influence to, that of many states and international organizations. As part of their emergence, civil society organizations began to exert pressure for recognition of their right to participate directly in the formation of global policy. And, indeed they did participate in a much larger (if still unofficial and indirect) way than in the past. For example, all of the big thematic United Nations conferences held during the 1990s (Environment (1992); Human Rights, (1993); Population (1994); Social Summit (1995); Women (1995)) included robust parallel proceedings attended by thousands of civil society representatives. These occasions gave civil society enhanced opportunities to influence inter-governmental debate and negotiations, engage in networking activities, and through accessing the media impact world public opinion. Symbolic of civil society's new status, the United Nations Secretary General capped this decade of participation in the year 2000 by inviting representatives of civil society to United Nations Headquarters for a Millennium NGO Forum, the purpose of which was to be an advisory forerunner to the Millennium Assembly of States.

Fortunately, the energy of global civil society could not be contained solely within the institutional confines established by global officialdom. Most dramatically, this activism spilled out into the streets of Seattle during the 1999 WTO Ministerial Conference when over 40,000 citizens protested the WTO's undemocratic procedures and policies in what came to be dubbed the *Battle of Seattle*. Other similar protests soon followed whenever and wherever those at the helm of the global system of economic governance were to meet, particularly the gatherings of the World Bank, WTO, IMF, and G-7.

As the first decade of economic globalization drew to a close in 2001, civil society itself channeled this energy into the founding of the World Social Forum as, at least in part, a symbolic counterweight to the World Economic Forum, the politically formidable and neoliberally oriented organization of business and political elites. In the years since the founding of the World Social Forum tens of thousands of representatives of diverse elements of civil society have met regularly in Porto Alegre and other non-Western cities in the hopes of advancing their varied agendas.

Paradoxically, despite all of the concern about global citizen participation voiced during this era, few observers gave consideration to whether a democratic role for citizens should and could be formally institutionalized within the international system. By 1997 we had come to conclude that thinking about global democracy had not kept pace with the new democratic and internationalist realities of the era. In particular, we believed that the times provided an opening for considering whether the central institution of national democracies — a popularly elected assembly or parliament — could be adapted to the global system. From our first *Op Eds* published in the popular press, to our later academic works, our goal for the writings contained in this book, has not been merely to provide a conceptual or normative analysis, but to contribute concretely to global democracy by making a politically compelling case for the institution of a popular assembly.

What had started as an ambitious proposal offered in 1997 during a time of great enthusiasm for democratization and internationalism became far more difficult to realize in 2001 following the contested election of George W. Bush as the new U.S. President and eight months later the attacks of September 11th. As al Qaeda became the most visible image of a non-governmental organization, the dramatic rise of civil society in the 1990s was to a great extent eclipsed by the revival of statist security concerns in the United States and elsewhere. Bowing to statist pressure, for example, the large participatory conferences held under UN auspices were largely phased out. As the reader will notice, our writings of this period respond to a political context where the loudest voices were those calling for what the Pentagon named the 'long war' and visions of global inclusion were for many people overcome by a Hobbesian mood of fear and tension, a dangerous development in a world where access to weapons of mass destruction was being universalized.

Though the last decade has been a difficult one for the global democracy movement, it has emerged with its fundamental goals and aspirations very much intact. Beneath the public radar of what may dominate the headlines of the day, the democratizing movement has continued to make impressive strides. In the academic world global democracy has become an important subfield of international relations and political theory. Leading scholars of cosmopolitan democracy such as Daniele Archibugi and David Held are cited far and wide,[1] and many others have contributed to working out the theory and practice of how application of democracy to the global order can extend beyond the liberal emphasis on elections and the rule of law.[2] During this period ideas have been developed, pondered over, and refined. For example, the reader of this volume will notice an evolution in our own thinking

---

[1] For some of their representative works, *see* Cosmopolitan Democracy: An Agenda for a New World Order (Daniele Archibugi & David Held eds., 1995); Daniele Archibugi, The Global Commonwealth of Citizens: Toward Cosmopolitan Democracy (2008).

[2] For one of many additional important works within the cosmopolitan democracy school *see,* Raffaele Marchetti, GLOBAL DEMOCRACY: FOR AND AGAINST: ETHICAL THEORY, INSTITUTIONAL DESIGN AND SOCIAL STRUGGLE (2008).

Many other scholars writing in the global democracy subfield have focused on the specific question of the extent to which intergovernmental organizations can be made more transparent and accountable to governments and other stakeholders. For two often cited and important works see, *e.g.,* Ruth W. Grant & Robert O. Keohane, *Accountability and Abuses of Power in World Politics*, 99 AM. POL. SCI. REV. 29, 29-43 (2005); Robert O. Keohane, *Global Governance and Democratic Accountability, in* TAMING GLOBALIZATION: FRONTIERS OF GOVERNANCE 130, 130-59 (David Held & Mathias Koenig-Archibugi eds., 2003).

Other scholars have focused on participation by civil society organizations and epistemic networks in the international system. For some of the more influential works on participation by civil society organizations see, MARGARET E. KECK AND KATHRYN SIKKINK, ACTIVIST BEYOND BORDERS (1998); Jessica T. Mathews, *Powershift*, FOREIGN AFF., JAN./FEB. 1997, at 50; JACKIE SMITH, SOCIAL MOVEMENTS FOR GLOBAL DEMOCRACY (2008). While the impact of civil society networks on global governance has been the topic of considerable discussion, Anne-Marie Slaughter's consideration of their melding with inter-governmental networks has probably been most influential. *See* ANNE-MARIE SLAUGHTER, A NEW WORLD ORDER (2004).

from our early endorsement of a parliament or assembly created by civil society to our later advocacy of an interstate treaty created parliament.

This is not to say that there yet exists an academic consensus supportive of the need for and desirability of a global parliament or even with respect to democratic global reforms in general. Andrew Moravcsik, for example, in the title of a provocative article asks the question, *Is There a 'Democratic Deficit' in World Politics?*[3] He concludes that, at least with regard to the European Union, the answer is in the negative, and other international specialists remain similarly skeptical that global democracy is emergent or a natural sequel to domestic democracy.[4] Though there may not be agreement about answers or even questions, it is certainly the case that the academic debate about global democracy has been joined and is likely to continue, and perhaps will even intensify.

Likewise, civil society has sustained its efforts to overcome the international system's democratic deficit, and a determined, if somewhat inchoate, movement for global democracy continues to evolve. Important players in that movement such as the Montreal International Forum,[5] founded in 1998, and Building Global Democracy,[6] founded in 2008, have established themselves as inclusive *big tents* by broadly defining their missions as advancing the cause of global democracy generally. In contrast, the focused purpose of the four year old Campaign for a United Nations Parliamentary Assembly[7] is to establish the institution of a parliament within the United Nations system. Under the leadership of its Secretary General, Andreas Bummel, the Campaign has obtained the backing of hundreds of parliamentarians from countries around the world.

Looking To The Future

As this book goes to press in the second decade of the 21st century the world is again changing, this time in ways that may portend future progress in the struggle for global democracy. As when we first began our parliament pro-

---

[3] Andrew Moravcsik, *Is there a 'Democratic Deficit' in World Politics?: A Framework for Analysis*, 39 GOV'T & OPPOSITION 336 (2004).

[4] *See* Robert O. Keohane, Stephen Macedo, Andrew Moravcsik, *Democracy-Enhancing Multilateralism,* 63 INT'L ORG. 1 (2009); Philip Pettit, *Two-Dimensional Democracy, National and International,* (N.Y.U. Sch. L. Inst. for Int'l L. & Just., Working Paper No. 2005/8, 2005), available at http://www.iilj.org/publications/2005-8Pettit.asp.

[5] http://www.fimcivilsociety.org/.

[6] http://www.buildingglobaldemocracy.org/.

[7] http://www.unpacampaign.org/.

ject in the 1990's, democracy is again on the march. This time, it is the inspiring Arab Spring revolts that stand as testament to an expanding consensus among the world's citizens that societies should be democratically constituted. Likewise, as in the 1990's, the trend is toward internationalism. In particular, in contrast to the period following 9/11, the United States is less likely to commit its oversized diplomatic weight to steadfastly opposing global solutions to the world's common problems. Perhaps of even more fundamental importance to the future of internationalism, the world seems headed toward an era where power will be more evenly balanced between the United States, the European Union and Japan and such rising regional and global actors as China, India, Brazil, Russia, and South Africa.

It is the meeting of these revitalized trends toward democracy and internationalism with the ever growing practical need for democratic reforms of the global order that gives us the greatest confidence that the future may be conducive to global democracy. While globalization continues to integrate the world's economies, the international system has exhibited its inability to respond well to the greatest period of financial instability since the Great Depression of the 1930s. And, even more distressing, existing institutions of global governance have shown themselves unable to deal effectively with either of the apocalyptic challenges of global warming and the possession and proliferation of nuclear and other weapons of mass destruction. While various interests may promote schemes to give the international system authoritarian powers to deal with these challenges, such schemes are not morally acceptable nor, for that matter, are they likely to be problem-solvers. In a world where the democratic spirit is increasingly taking hold, citizens are not likely (short of force) to accept despotically imposed solutions to contentious global issues.

By this assessment we do not wish to understate the significant obstacles that remain to galvanizing a critical mass of support for a global parliament. Certainly, many powerful institutions of the global order perceive such a body as contrary to their interests. Even certain civil society organizations (including some proponents of increased civil society access to global institutions) seem to perceive a need to guard jealously their claim to the mantle of *voice of the global citizenry* from being taken over by the seemingly superior claim to representativeness of a popularly elected body. Likewise, many democratic governments seem far from enthusiastic about relinquishing a portion of their current control over global political institutions and procedures to a popularly elected chamber that might favor policies that are at odds with their own preferences. Perhaps most paradoxical are the evolving attitudes toward a global parliament by the world's non-democracies. Certainly most absolutist rulers would rather not be forced into choosing between the reputational costs of precluding their citizens from participating in global elections and the political threat of acceding to the introduction of democratic practices into their countries. On the other hand, many of the world's constitutional oligarchies, such as China, appear to be gradually warming to the possibility that a globally representative institution could help break the West's disproportionate influence over global institutions without necessarily threatening their internal political structures.

Beyond the difficulty of negotiating the labyrinth of formidable institutional interests is the challenge of popular appeal. Can the theoretical case for a global parliament be formulated so that it resonates widely with the peoples of the world and their leaders? To be sure the nature of the project does not easily lend itself to a simple mobilizing message such as that put out by the Coalition for an International Criminal Court's: *the Milosevics of the world should not be allowed to get away with it,* or the Campaign to Ban Landmines', *children should not lose their limbs to landmines.* In addition, any message with a hope of reaching a large audience must be responsive to the worldview of a global public that vacillates between the poles of universalism and tribalism, and that, while generally preferring democracy to tyranny, is often made cynical about electoral politics by corruption in domestic democracies and by economic priorities that do not favor the common person. Finally, many in positions of global influence who understand the need for global democratic reform and a global parliament have not freed their political behavior from deference to short-term priorities, believing that their future depends on resolving current crises, rather than bringing about long term structural change.

Despite these challenges, we continue to believe as strongly in the democracy project to which this book is devoted as we did when we began in 1997. The proposal for a global parliament — as we emphasize repeatedly in the chapters that follow — is not utopian. Indeed, since the 18th Century the forward march of the democracy idea has consistently beat the odds in transforming country after country. There is no law of political nature which says democratic change cannot come as well to the international system. Today, as the publishers of this book, the Committee for a Democratic United Nations, and the International Campaign for a UN Parliamentary Assembly stand as testament to, much of the organizational infrastructure is in place. What awaits is for a critical mass of those who continue to believe in the twin values of democracy and internationalism to mobilize behind this project. It is to these as yet dormant persons that we address this book in the fervent hope of awakening their latent sense of mission.

1
On Globalization, Democracy, and the Need for a GPA

Toward Global Parliament

by Richard Falk and Andrew Strauss

Foreign Affairs, 2001[*]

Challenging the Democratic Deficit

One crucial aspect of the rising disaffection with globalization is the lack of citizen participation in the global institutions that shape people's daily lives. This public frustration is deeper and broader than the recent street demonstrations in Seattle and Prague. Social commentators and leaders of citizens' and intergovernmental organizations are increasingly taking heed. Over the past 18 months, President Clinton has joined with the secretary-general of the United Nations, the director-general of the World Trade Organization (WTO), the managing director of the International Monetary Fund (IMF), and the president of the World Bank to call for greater citizen participation in the international order.

But to date, these parties have not clearly articulated a general vision of how best to integrate a public role into international institutions. So in the absence of a planned design, attempts to democratize the international system have been ad hoc, as citizen organizations and economic elites create their own mechanisms of influence. In domestic politics, interest-group pluralism flourishes within a parliamentary system of representation. In global politics, interest-group pluralism is growing, but no unifying parliament represents the public interest. This state of affairs cannot last in a world where the prevailing understanding of democracy does not accept the fact that unelected interest groups can speak for the citizenry as a whole. Any serious attempt to challenge the democratic deficit must therefore consider creating some type of popularly elected global body. Before globalization, such an idea would have been considered utopian. Now, the clamor of citizens to participate internationally can no longer be ignored. The only question is what form this participation will take.

[*] Reprinted by permission of FOREIGN AFFAIRS, Volume 80, No. 1, January/February 2001. Copyright (2001) by the Council on Foreign Relations, Inc.

Decision-making goes global

Behind this clamor lies a profound shift in power. Thanks to trade, foreign direct investment, and capital flows, globalization is dispersing political authority throughout the international order. International governance is no longer limited to such traditional fare as defining international borders, protecting diplomats, and proscribing the use of force. Many issues of global policy that directly affect citizens are now being shaped by the international system. Workers can lose their jobs as a result of decisions made at the WTO or within regional trade regimes. Consumers must contend with a market in which state-prescribed protections such as the European ban on hormone-fed beef can be overridden by WTO regulations. Patients who need medicines pay prices influenced by WTO-enforced patent rules, which allow pharmaceutical companies to monopolize drug pricing. Most of the 23 million sub-Saharan Africans who have tested positive for the AIDS virus cannot afford the drugs most effective in treating their illness. They will die much sooner as a consequence.

For the half of the world's population that lives on less than $2 a day, governmental social safety nets have been weakened by IMF decisions. The globalized economy has not meaningfully reduced poverty despite a long period of sustained growth. Economic inequality is on the rise, as is the marginalization of regions not perceived as attractive trading partners or "efficient" recipients of investment. Furthermore, environmental trends pose severe dangers that can be successfully dealt with only through global action and treaties. Against such a background, it is little wonder that people who believe they possess a democratic entitlement to participate in decisions that affect their lives are now starting to demand their say in the international system. And global civil society has thus far been their voice as they attempt to have this say.

Civil society's global presence

Civil society, made up of nonprofit organizations and voluntary associations dedicated to civic, cultural, humanitarian, and social causes, has begun to act as an independent international force. The largest and most prominent of these organizations include Amnesty International, Greenpeace, Oxfam, and the International Committee of the Red Cross; in addition, the U.N. now lists more than 3,000 civil society groups.

During the 1990s, these transnational forces effectively promoted treaties to limit global warming, establish an international criminal court, and outlaw antipersonnel land mines. These same actors also helped persuade the International Court of Justice to render an advisory opinion on the legality of

nuclear weapons and defeat a multilateral investment agreement. More recently, civil groups mounted a drive to cancel the foreign debts of the
world's poorest countries. Although these efforts remain works in progress,
civil society to date has been indispensable in furthering them.

During the early 1990s, civil society's organizations began visibly coop-
erating at large international conferences of states. When conservative political pressures forced an end to these conferences, civil society began to coalesce to act cohesively and independently in the international arena. For example, 8,000 individuals representing civil society organizations met in May
1999 at the Hague Appeal for Peace to shape strategy and agree on a common agenda. Among those attending were such luminaries as Nobel Peace
Prize winners Desmond Tutu, Jose Ramos-Horta, and Jody Williams. Similar
smaller meetings in South Korea, Canada, Germany, and elsewhere followed.

These meetings were a prelude to the Millennium NGO Forum held at the
United Nations in May 2000, to which U.N. Secretary General Kofi Annan
invited 1,400 individuals representing international civil society groups to
present views on global issues and citizen participation in decision-making.
The forum agreed to establish a permanent assembly of civil society organizations, mandated to meet at least every two to three years, before the U.N.
General Assembly annual session. Although it is still to be realized, such a
forum might earn recognition over time as an important barometer of world
public opinion — and a preliminary step toward creating a global parliament.
Regardless of how this specific forum develops, civil society will continue to
institutionalize itself into an independent and cohesive force within the international system.

The corporate movers

Through expanding trade and investment, business and banking leaders have
also exercised extraordinary influence on global policy. Even in formerly
exclusive arenas of state action, these private-sector actors are making a
mark. For example, Secretary-General Annan has made "partnering" with the
business community a major hallmark of his leadership. The United Nations
has now established a formal business advisory council to formalize a permanent relationship between the corporate community and the U.N.

As with citizen groups, elite business participation in the international
system is becoming institutionalized. The best example is the World Economic Forum in Davos, Switzerland. In the 1980s, the WEF transformed
itself from an organization devoted to humdrum management issues into a
dynamic political forum. Once a year, a thousand of the world's most power-

ful business executives get together with another thousand of the world's senior policymakers to participate in a week of roundtables and presentations. The WEF also provides ongoing arenas for discussion and recommendations on shaping global policy. It is notable that Annan's ideas about a U.N. partnership with the business community have been put forward and endorsed during his frequent appearances at Davos. In addition, the WEF also conducts and disseminates its own research, which not surprisingly shows a consistently neoliberal outlook. For example, it produces a well-publicized annual index ranking the relative economic competitiveness of all countries in the world. The Davos assembly and overlapping networks of corporate elites, such as the International Chamber of Commerce, have been successful in shaping compatible global policies. Their success has come in the expansion of international trade regimes, the modest regulation of capital markets, the dominance of neoliberal market philosophy, and the supportive collaboration of most governments, especially those of rich countries.

Pondering a global parliament

Global civil society still cannot match the resources and power linkages of the corporate and banking communities. But many civil society groups have carved out niches within the international order from which to influence decision-making by relying on imagination and information. The evolution of these two networks — civil and business — has been largely uncoordinated, and it remains unclear how they could fit together in a functionally coherent and representative form of global governance. Neither can claim to represent citizenry as a whole. As global civil society acquires a greater international presence, its critics are already challenging its claims to represent the public interest. The charge of illegitimacy has even greater resonance when leveled at corporate and banking elites, who do not speak for organizations.

Now that the global system is increasingly held up to democratic standards — and often comes up short — those people who find their policy preferences rejected are unlikely to accept the system's determination as legitimate, and the democratic deficit will remain a problem. Only when citizen and business interests work together within an overarching representative body can they achieve policy accommodations that will be seen as legitimate. For the first time, a widely recognized global democratic forum could consider environmental and labor standards and deliberate on economic justice from the perspectives of both North and South. Even an initially weak assembly could offer some democratic oversight of international organizations such as the IMF, the WTO, and the World Bank.

Unlike the United Nations, this assembly would not be constituted by states. Because its authority would come directly from the global citizenry, it could refute the claim that states are bound only by laws to which they give their consent. Henceforth, the ability to opt out of collective efforts to protect the environment, control or eliminate weapons, safeguard human rights, or otherwise protect the global community could be challenged.

In addition, the assembly could encourage compliance with established international norms and standards, especially in human rights. The international system currently lacks reliable mechanisms to implement many of its laws. Organizations such as Amnesty International, Human Rights Watch, and even the International Labor Organization attempt to hold states accountable by exposing their failures of compliance, relying on a process often referred to as the "mobilization of shame." In exercising such oversight, a popularly elected global assembly would be more visible and credible than are existing watchdogs who expose corporate and governmental wrongdoing.

The assembly's very existence would also help promote the peaceful resolution of international conflicts. Because elected delegates would represent individuals and society instead of states, they would not have to vote along national lines. Coalitions would likely form on other bases, such as world-view, political orientation, and interests. Compromises among such competing but nonmilitarized coalitions might eventually undermine reliance on the current war system, in which international decisions are still made by heavily armed nations that are poised to destroy one another. In due course, international relations might more closely resemble policymaking within the most democratic societies of the world.

All those in favor

In spite of its advantages, would the formation of such an assembly threaten established state and business interests so much that its creation would become politically untenable? The European Union's experience suggests otherwise. Established by states — and with little initial authority — the transnationally elected European Parliament has now become powerful enough to help close a regional democratic deficit.

As with the early European parliament, a relatively weak assembly initially equipped with largely advisory powers could begin to address concerns about the democratic deficit while posing only a long-term threat to the realities of state power. Systemic transformation of world order that would largely affect successors would not significantly threaten those political leaders who are inclined to embrace democratic ideals. Indeed, it might even appeal to them.

Despite these humble origins, the assembly would have the potential to become an extremely important fixture of the global architecture. Upon the assembly's inception, civil society organizations would almost certainly lobby it to issue supportive resolutions. Groups who opposed such resolutions could shun the process, but that is not likely: they would concede the support of the world's only elected democratic body. Over time, as the assembly became the practical place for clashing interests to resolve differences, formal powers would likely follow.

Some business leaders would certainly oppose a global parliament because it would broaden popular decision-making and likely press for transnational regulations. But others are coming to believe that the democratic deficit must be closed by some sort of stakeholder accommodation. After all, many members of the managerial class who were initially hostile to such reform came to realize that the New Deal — or its social-democratic equivalent in Europe — was necessary to save capitalism. Many business leaders today similarly agree that democratization is necessary to make globalization politically acceptable throughout the world.

As the recent large street protests suggested, globalization has yet to achieve grassroots acceptance and legitimacy. To date, its main claim to popular support is not political but economic: it has either delivered or convincingly promised to deliver the economic goods to enough people to keep the anti-globalization forces from mounting an effective challenge. But economic legitimacy alone can rarely stabilize a political system for long. Market-based economic systems have historically undergone ups and downs, particularly when first forming. The financial crisis that almost triggered a world financial meltdown a few years ago will not be the last crisis to emerge out of globalization. Future economic failures are certain to generate political responses. Standing in the wings in the United States and elsewhere are politicians, ultranationalists, and an array of opportunists on both the left and the right who, if given an opening, would seek to dismantle the global system. A global parliament is therefore likely to serve as an attractive alternative to those people who, out of enlightened self-interest or even public spiritedness, wish to see the international system become more open and democratic.

Making it happen

Although the raw political potential for a global assembly may exist, it is not enough. Some viable way needs to be found for this potential to be realized, and it can most likely be found in the new diplomacy. Unlike traditional diplomacy, which has been solely an affair among states, new diplomacy makes room for flexible and innovative coalitions between civil society and receptive states. The major success stories of global civil society in the 1990s — the Kyoto global warming treaty, the convention banning land mines, and the International Criminal Court — were produced in this manner.

Civil society, aided by receptive states, could create the assembly without resorting to a formal treaty process. Under this approach, the assembly would not be formally sanctioned by states, so governments would probably contest its legitimacy at the outset. But this opposition could be neutralized to some extent by widespread grassroots and media endorsement. Citizens in favor could make their voices heard through popular, fair, and serious elections.

Another approach would rely on a treaty, using what is often called the "single negotiating text method." After consultations with sympathetic parties from civil society, business, and nation-states, an organizing committee could generate the text of a proposed treaty establishing an assembly. This text could serve as the basis for negotiations. Civil society could then organize a public relations campaign and persuade states (through compromise if necessary) to sign the treaty. As in the process that ultimately led to the land mines convention, a small core group of supportive states could lead the way. But unlike that treaty, which required 40 countries to ratify it before taking effect, a relatively small number of countries (say, 20) could provide the founding basis for such an assembly. This number is only a fraction of what would be needed for the assembly to have some claim to global democratic legitimacy. But once the assembly became operational, the task of gaining additional state members would likely become easier. A concrete organization would then exist that citizens could urge their governments to join. As more states joined, pressure would grow on nonmember states to participate. The assembly would be incorporated into the evolving international constitutional order. If it gained members and influence over time, as expected, its formal powers would have to be redefined. It would also have to work out its relationship with the U.N. One possibility would be to associate with the General Assembly to form a bicameral world legislature.

The pressures to democratize the international system are part of an evolutionary social process that will persist and intensify. The two dominant themes of the post-Cold War years are globalization and democratization. It is often said that the world is rapidly creating an integrated global political economy, and that national governments that are not freely elected lack political legitimacy. It is paradoxical, then, that a global debate has not emerged on resolving the contradiction between a commitment to democracy and an undemocratic global order.

This tension may be the result of political inertia or a residual belief that ambitious world-governance proposals are utopian. But whatever the explanation, this contradiction is spurring citizen groups and business and financial elites to take direct actions to realize their aspirations. Their initiatives have created an autonomous dynamic of ad hoc democratization. As this process continues to move along with globalization, pressures for a coherent democratic system of global governance will intensify. Political leaders will find it more difficult to win citizen acquiescence to unaccountable policies that extend globalization's reach into peoples' lives. To all those concerned about social justice and the creation of a humane global order, a democratic alternative to an ossified, state-centered system is becoming ever more compelling.

Globalization Needs a Dose of Democracy

by Richard Falk and Andrew Strauss

The International Herald Tribune, 1999[*]

T he economic and political problems of the last year have driven home just how reliant the world has become upon effective international solutions to what would previously have been considered regional or local problems.

A foreign reserve shortfall in Thailand triggered an economic crisis in emerging markets that very possibly would have engulfed the whole of the world economy if not for extensive intervention by the IMF.

Human rights crises in tiny Kosovo and in East Timor were seen as having profoundly destabilizing implications and as calling for significant military responses by NATO and the United Nations.

Because of the urgent demands of a more interconnected, globalized world it seems inevitable that the international order will play a significantly enhanced role in the next century. There is, however, no structure in place to ensure that this order will be organized along democratic lines.

In fact, despite the increasing importance placed on democratization of domestic governance, almost no attention has been devoted to ensuring that this most fundamental political value is applied to the increasingly important global dimensions of our politics.

Accepting the challenge to extend democracy beyond its familiar link to the state does not tell us how this might best be done. We believe that the most promising innovation would be a worldwide grassroots campaign to establish the first Global Peoples' Assembly.

To many this idea must seem fanciful. Certainly it seems unlikely that most governments would support a proposal that would threaten their monopoly in the global arena. But governmental reluctance need not be decisive.

Globalization is creating a nongovernmental global civil society composed of nongovernmental organizations, transnational business, labor, media, cultural and religious institutions and networks, and cosmopolitan individuals with extraordinary wealth and influence. This numerically small yet highly visible globalized citizenry now has the capacity, perhaps with the help of some forward-looking governments, to organize such an assembly.

If, as the democratic principle asserts, political authority ultimately resides in citizens, then the citizenry has the right to found its own assembly.

How could we proceed to bring this assembly into being? Perhaps the most effective initial move would be to issue an appeal endorsed by moral authority figures (religious leaders, Nobel Peace Prize laureates) that calls on the peoples of the world to bring about such an assembly. If well-executed, this appeal would probably succeed in raising needed organizing funds.

As a second stage, meetings could be arranged throughout the world with the goal of forming a citizens' committee that could organize and administer global elections. A voting formula based upon one person, one vote would probably be acceptable and fairest. Elections could then be held, monitored by respected observers.

Along the way many stumbling blocks would of course arise. Global voter roles would have to be generated. A system of campaign finance and other election rules would need to be established, and attempts to manipulate or undermine elections would have to be effectively guarded against.

Some governments would undoubtedly not allow elections to occur in their countries. As a result, these societies would initially be unrepresented, or temporarily represented by a selection process carried out among citizens in exile.

There is little reason to believe that logistical and political problems could not be solved by sufficiently dedicated participants. The innovation of a global assembly seems far less radical than was the bitter historical struggle waged for centuries against royal absolutism to establish parliamentary bodies representative of the citizenry.

Once established, the assembly would play a modest role at first. But if it achieved a respected presence over a period of years, it would begin to influence governments and media. At this point it could seek formal inclusion within the United Nations system.

Until this point is reached, the assembly would have an international legal status similar to that of such nongovernmental organizations as the Red Cross, Amnesty International or the International Olympic Committee — but with one big difference. It could lay some claim to represent all the peoples of the world.

As the only such body, it would have the potential to become influential long before receiving formal recognition. Parliament in England, after all, began as an informal advisory body whose influence as the sole representative of the people gradually achieved potency.

In our own time, the increasingly powerful, directly elected European Parliament existed for many years as a largely symbolic representative of the peoples of the European Union.

The global assembly could usefully contribute to the creation of planetary norms by expressing views on critical issues of global policy. Not only could such an assembly be a vehicle for championing social justice, it could greatly contribute to the development of a more peaceful global order.

Representatives from different countries and civilizations would convene in a climate of civility to advance mutual interests and discuss differences. Interest groups trying to influence the assembly would coalesce across national lines.

The normal parliamentary business of delegates working to build social consensus on issues would encourage the development of universal values over more parochial concerns and beliefs.

A more democratic world, in which individuals and groups are less likely to perceive their rights as threatened, is a world more likely to be at peace.

The rise of global civil society at this auspicious dawning of a new millennium gives us the opportunity to participate in creating a democratic system in which governmental power, domestic and international, is derived directly from the consent of the governed.

Given a global peoples' assembly, the grander project of a democratic form of global governance could proceed with considerable confidence.

International Law as Democratic Law

by Andrew Strauss

American Society of International Law Proceedings, 2010*

Normative political theorists have traditionally seen democracy as limited to domestic governance. When Plato and Aristotle attempted to envisage superior forms of government, they assumed that the realm relevant to their consideration was the domestic. Likewise, when James Madison conceived of the American republican or democratic constitution, he assumed that it was to bring about a *more perfect* internal union. Throughout the 20th and now 21st century, however, power has continued to evolve to the international system. Today we not only have international law, but we have rapidly proliferating and increasingly empowered international organizations which have institutionalized global governance.

Given these developments, theorists in both international law and international relations are increasingly noting the anomaly of a normative perspective which holds that democracy is the *sin qua non* of legitimate domestic governance but disregards its importance in the international system. Despite this attention paid to the global democratic deficit, however, precisely in what way the international system fails to be democratic remains undertheorized. This is troubling because it is only with a clear conceptual framework for thinking about the problem that we can understand its dimensions and evaluate various responses to it.

This afternoon I, therefore, wish to start with the proposition that provision for the formal political equality of citizens is among the basic criteria—not the only—but among the basic criteria a social order must meet to be considered democratic. As Robert Post has succinctly put it in the *Annals of the American Academy of Political and Social Sciences*, "Democracy requires that persons be treated equally insofar as they are autonomous agents participating in the process of self-government. This form of equality is

* Reprinted from PROCEEDINGS OF THE 103 ANNUAL MEETING OF THE AMERICAN SOCIETY OF INTERNATIONAL LAW, 2010.

foundational to democracy, because it follows from the very definition of democracy."[8]

To what extent then does the structure of the international system allow for such political equality? Drawing, by way of comparison, on Robert Dahl's notions of interest group pluralism,[9] we can usefully summarize that within modern representative democracies this equality is institutionalized in the granting of every citizen a theoretically equal opportunity to influence political outcomes. The ultimate arbiters of governmental policy are representatives selected by citizens who each have an equal vote or say in that selection. After this selection citizens continue to influence representatives by way of interest groups which all citizens have a formally equal opportunity to participate in forming. Such interest group formation is dynamic in that citizens can freely shift their allegiances among interest groups and the causes they represent. While this formal equality of opportunity to influence political outcomes is for a variety of reasons only imperfectly realized in even the most successful national democracies, these democracies are nevertheless structured so as to approximate the ideal.

The international system, on the other hand, is structured so as to preclude the ideal of political equality. Citizens do not have a formally equal opportunity to select representatives to be the ultimate arbiters of policy in the international system. Rather, vastly unequal states themselves perform this role, and in the terms of our discussion of interests groups, all of the citizens of a certain nationality and/or within a certain geographical area are part of a permanent political coalition frozen into the institutional mold of the state. If citizens disagree with a political position their state is taking in the international system, they cannot freely shift their support to other more like-minded states as they would interest groups within a domestic democratic arena. Rather their state's command over their economic resources (through taxes and their contribution to the state's economy) and often human resources (through conscription) as well as enforced conformity with the state's legal edicts all result in their having no choice but to contribute to the state's position.

This has significant implications for democratic equality. Citizens do not enjoy theoretical equality of opportunity to influence political outcomes in the international system. The influence of citizens from powerful states who prevail internally in getting their states to promote their external policy preferences is magnified by the power of the state while those who do not prevail

---

[8] Robert Post, *Democracy and Equality,* 603 ANNALS OF THE AM. ACAD. OF POL. AND SOC. SCI., 24, 28 (2006).

[9] *See, e.g.* ROBERT DAHL, DEMOCRACY AND ITS CRITICS (1989).

are given no official voice in the state centric process of making global policy.

As problematic as this might be, citizens of strong states that are internally democratic at least have a theoretical equality of opportunity to prevail in getting their states to promote their external policy preferences. Citizens of weak states, on the other hand, are restricted in their ability to influence global decision-making even when they prevail internally because their states themselves are so restricted. The bottom line is that fairly small numbers of citizens from powerful countries who represent far less than a global consensus can leverage their position so as to allow them to prevail in global decision-making.

This is not merely an abstract problem. All that I am doing this afternoon through theory is offering an analytical framework for understanding what millions of people around the world (and I'm sure many colleagues in this room) already know. One has only to travel to global civil society conferences, read relevant public opinion polls or even the headlines of daily newspapers to understand that there is a profound feeling among many people that they are not being fairly represented in global decision-making.

The good news is that this mass dissatisfaction can provide the raw fuel necessary—if properly channeled—to energize institutional innovation. How long after all can the world continue to democratize and to globalize without the two coming together? Interpreting the normative in the title of our panel as meaning prescriptive let me then *prescribe* what I see as the only serious way to overcome the global lack of political equality—and that is democracy. Specifically, I wish to suggest the initiation of what has come to be regarded in democracies as the core democratic institution: a parliament.

Richard Falk and I have explored elsewhere strategies for creating a popularly elected global parliament modeled on the European Parliament, and a discussion of such strategies is beyond the subject of this talk. But suffice to say that the path to a parliament may not be as difficult as many people assume, particularly if the parliament starts as an advisory body and with less than universal membership.[10]

The central question my fellow panelists are grappling with is how to identify a solid theoretical basis for international law as law. A great deal is at stake in the endeavor, not only normatively, but practically. To the extent such a basis gains widespread acknowledgement, international law becomes more legitimate and hence more effective. My contribution then is to introduce into the discussion an additional basis for international (or perhaps now *trans*national) law—democracy.

---

[10] *See, e.g.* Richard Falk & Andrew Strauss. *Toward Global Parliament,* 80 FOREIGN AFF. 212 (2001).

Rather than trying to just make states subject to their own collective law, I am suggesting that for certain purposes we transcend states to rest global authority upon *the consent of the governed.* This well established democratic principle would ground global law in the same basis of authority as domestic democratic law. And over time global democracy would hold out at least the possibility of creating similarly effective global institutions.

Certainly, there are many questions regarding how a global parliament would function and its relationship to other institutions of global governance, but a discussion of those must necessarily be left for another time.[11] It is, however, also relevant to question whether an institutional innovation as significant as a global parliament is necessary to introduce the ideal of political equality to the global system. Let me then conclude by briefly critiquing what I see as the three major non-parliamentary alternatives that are being offered for making the global system more democratic.

The first suggested by commentators such as Jeffrey Laurenti is to promote democratization at the national level.[12] Arguably if freely elected national leaders are capable of democratically representing their constituents domestically, then shouldn't they likewise also be capable of representing them internationally? I have shown, however, that the global system distorts this representation. States, as vastly unequal "representatives," are not the same as similarly situated parliamentarians whose constituents all have a theoretically equal say in their selection. Likewise, states in their role as "frozen" interest groups do not give all citizens a theoretically equal opportunity to influence political outcomes. Domestic democratization without more, therefore, will not allow for all citizens to have formal equality of political opportunity within the international system.

The second alternative suggested by many commentators is to expand the ability of interest groups to participate as members of *transnational epistemic networks.*[13] These commentators variously suggest that interest groups more often be granted observer status in international organizations, that they be more liberally allowed to submit amicus briefs to international tribunals and policy recommendations to treaty bodies. There is no doubt that citizens can bypass their states to play some role in directly affecting decision-making in international fora, But ultimately the decision-makers are not democratic representatives operating in a fundamentally pluralist system but

[11] *See, Id.*

[12] *See,* Jeffrey Laurenti, *An Idea Whose Time Has Not Come, in* A READER IN SECOND ASSEMBLY & PARLIAMENTARY PROPOSALS 119, 128-129 (Saul H. Mendlovitz and Barbara Walker eds., 2003).

[13] *See, e.g.* essays in GLOBAL CITIZEN ACTION (Michael Edwards & John Gaventa eds., 2001).

states operating on a vastly unequal playing field where non-state actors have no structural role and their influence is therefore destined to be at the margins.

The final alternative suggested by Joseph Nye[14] among many other is to provide for greater accountability and transparency of international organizations. Institutional reforms that help overcome the lack of democratic transparency and accountability within the global system are no doubt valuable. They do not, however, address the lack of equality of opportunity for citizens to influence global outcomes.

Thus, I do not think that present global institutions can be tinkered with to achieve political equality. I believe we need to consider fundamental reform, and this I believe is the great hidden civil rights challenge of our time. But for the analytical blinders which direct our view of global politics exclusively to the interplay between sovereign states, it would be clear to everyone that the global system does not measure up to generally professed democratic values. Faced with great dissatisfaction and the resulting political discord, for the powerful to choose coercion over reform as has often been the case is to betray democratic values and diminish the hope of a global system which can be sustained through legitimacy rather than force. As with our own American civil rights history, this is a time that offers a choice between inviting people in and forcibly keeping people out.

---

[14] Joseph S. Nye, *Globalization's Democratic Deficit: How to Make International Institutions More Accountable,* FOREIGN AFF. (July/August 2001).

Should Citizens Be Democratically Represented In the 21st Century International System?

by Andrew Strauss

ILSA Journal of International & Comparative Law, 2010[*]

I. Introduction

Democracy is increasingly the *sine quo non* of legitimate governance at the local, provincial and national levels. As power increasingly transfers to the international system, a decent collective future for humankind is likely to hinge on the extent to which we can integrate the values of democracy into the global system. If a more than cosmetic democratization of the global order is to occur, some institution or institutions representing citizens directly will need to take a prominent place within it.

Different people will have different ideas about how this can be done, and I do not wish to suggest that there is only one way parliamentary processes can be institutionalized at the global level.[1] But to make today's discussion concrete, and to encourage further dialogue, I will first present the proposal for a global democratic organization that Richard Falk and I have been promoting.[2] I will then discuss why I believe some initiative along the lines we are suggesting is important to our collective future.

---

[*] Reprint from ILSA JOURNAL OF INTERNATIONAL & COMPARATIVE LAW, Volume 16, 2010.

[1] For example I compare four different approaches to initiating a global parliament in Andrew Strauss, *On the First Branch of Global Governance*, 13 WIDENER L. REV. 347 (2007).

[2] For some of our representative works see, for example, Richard Falk and Andrew Strauss, *Toward Global Parliament*, 80 FOREIGN AFF. 212 (January/February 2001); Richard Falk and Andrew Strauss, *On the Creation of a Global Peoples Assembly: Legitimacy and the Power of Popular Sovereignty*, 36 STAN J. INT'L L. 191 (2000); Richard Falk and Andrew Strauss, The Deeper Challenges of Global Terrorism, A Democratizing Response, *in* DEBATING COSMOPOLITICS, (Archibugi, Ed., 2003).

II. The Proposal for a Global Parliamentary Assembly

Richard Falk and I have been arguing that it is time to respond to the crisis of democratic legitimacy with the creation of a Global Parliamentary Assembly (GPA), a popularly elected world parliamentary body modeled on the European Parliament. The key to this proposal for a global parliament is that it is potentially transformative of the global system, while at the same time being politically realistic. It is potentially transformative in that its ultimate goal is a universally elected world body with limited but important legislative powers.[3] It is politically realistic in that it prescribes a process of incremental steps for bringing this vision to fruition. We propose that the GPA be constituted by a stand alone treaty agreed to by a vanguard of progressive democratic countries.[4] We have suggested that even as few as twenty geographically and economically diverse countries could be enough to found the GPA.[5] At the beginning, so as not to be too threatening to existing national leaders, the GPA's powers could be largely precatory. A civil society campaign could help create a political climate conducive to the successful conclusion of such a treaty.[6] The hope would be that the success of the kinds of initiatives best exemplified by the creation of the International Criminal Court could be repeated.

One lesson of the last half of the twentieth century is that Promethean plans to instantly transform global governance are exceedingly difficult to realize. What has proved successful, however, is the incrementalism that culminated in the European Union. Though the future cannot be predicted, once in place a GPA would be poised to grow in influence. Citizen elections would give it a unique visibility, and as the only global institution with a popular mandate, citizen groups would likely petition the parliament to pass

---

[3] The scope of the GPA's ultimate powers would, of course, be the subject of wide ranging negotiations involving multiple stake holders over an extended period of time. There is, however, likely to be strong consensus on two limitations to the GPA's powers as fundamental to its status as a globally democratic body. The first is that its prerogatives be limited by the human rights protections enshrined in the Universal Declaration of Human Rights and other similar human rights instruments. The second is that the GPA's powers be limited by the principle of subsidiary, that it should only legislate in areas that are appropriately in the international, as opposed to the domestic, sphere.

[4] Richard Falk and Andrew Strauss, *Toward Global Parliament*, 80 FOREIGN AFF. 212, 219 (January/February 2001).

[5] *Id.*

[6] Such an effort, The Campaign for a United Nations Parliamentary Assembly, was launched in 2007 and has succeeded in garnering support for the parliamentary idea from many different corners of the international community. For example, close to 700 members of parliament in 94 countries have signed the Campaign's appeal. *See,* http://en.unpa-campaign.org/about/index.php.

resolutions supporting their causes. Those opposed, be they industrial lobbies, states or other citizen groups, would be unlikely to willingly cede the parliament's popular legitimacy to their policy opponents. Instead, playing out the familiar process of parliamentary politics, they would likely participate as well. The parliament's arena could grow as a much needed venue where the various global interests could directly interact to promote their positions and resolve differences without having to rely on their respective governments to be intermediaries.

Once the GPA was established, citizen groups from countries the world over could petition their governments to join, and once a critical mass of membership was reached even authoritarian governments would find it costly politically to deny their citizens the right to be represented through free and fair elections in the one globally democratic institution. At some point in its evolution the Parliament's formal legal powers, as well as its relationship with the United Nations, would have to be settled. Perhaps it could, along with the General Assembly, be a part of a bicameral global legislative system.

III. The Case for a Global Parliamentary Assembly

A. Democracy

The international system is not organized along democratic lines. Most significantly, it does not provide citizens, or even states, an equal democratic right to participate in the political process. The UN Security Council, for example, does not allow meaningful citizen participation, and it only includes representatives from 15 countries.[7] As is well known, even delegates from most of those 15 countries are not fully enfranchised as the Security Council's five permanent members can unilaterally veto non procedural resolutions.[8] Even organizations that are ostensibly more democratic such as the World Trade Organization (where voting is based on member consensus)[9] are in truth largely controlled by the dictates of a few dominant members.[10]

---

[7] U.N. Charter art. 23, para. 1.

[8] *Id.* art 27, para. 3.

[9] Article IX of the Agreement Establishing the World Trade Organization provides that, the World Trade Organization "shall continue the practice of decision-making by consensus " Under Article IX voting is, however, allowed in the event consensus cannot be reached, but resort to this provision has been very rare. *See* Marrakesh Agreement Establishing the World Trade Organization, art. IX para. 1, 33 I.L.M. 1140 (1994).

[10] *See* Sonia E. Rolland, *Developing Country Coalitions at the WTO: In Search of Legal Support*, 48 HARV. INT'L L.J. 483 (2007) (discussing the practice of 'Green Room" decision-

As the demands of globalization increasingly transfer power from many relatively democratic national systems to the undemocratic international system,
the future of democracy may depend upon the international system being
democratized.

A GPA would introduce democratic equality into the global system. If
democratically constituted, it would grant citizens a substantially equal voice
in choosing their representatives, and all citizens would have access to the
parliament's processes. In addition a GPA would further the goal of making
international organizations democratically accountable to the global citizenry. While the powers of the GPA would grow gradually, even from its inception, it could play a positive advisory role in democratically overseeing the
global system by holding hearings, issuing reports and passing resolutions.
To have the Director General of the World Trade Organization, for example,
appear before the only popularly elected global body to answer to citizen
representatives would introduce some popular accountability into the system.

B. Effective Global Governance

The undemocratic international system is often unable to enforce law on
states. This means that it cannot effectively act to protect vital community
interests, such as in the control and elimination of weapons, the preservation
of the earth's biosphere, and the protection of fundamental human rights. The
problem is that to maximize their autonomy, governments have erected a
global system where with limited exceptions a state is bound only by laws it
agrees to,[11] and even once bound, states often flaunt those laws they find
disagreeable or inconvenient.

making whereby important decisions are made outside of formal processes by the most influential countries.)

[11] The two primary sources of international law are treaties and customary international
law. States maintain that they are only bound by treaties to which they agree to become party.
States, likewise, generally maintain that they are not bound to general customary international
law if they object to the custom at the time of its formation (the persistent objector rule). *See*
Luigi Condorelli, *Custom, in* International Law: Achievements and Prospects 179, 205 (Mohammed Bedjaoui ed., 1991). Because states can seldom demonstrate that they manifested this
lack of consent, customary international law is not completely consensual. *See generally,* J.
Patrick Kelly, *The Twilight of Customary International Law,* 40 Va. J. Int'l L. 449 (2000). In
addition, most states accept that states cannot opt out of a certain limited class of international
jus cogens norms that are so fundamental to international society's core values as to be non
derigible. *See, e.g.*, Vienna Convention on the Law of Treaties, May 23, 1969, art. 53, 1155
U.N.T.S. 331, 334 (1969) (providing that a treaty is "void if … it conflicts with a peremptory
norm of general international law"). Peremptory norms are probably inclusive of core human
rights standards. *See* Lauri Hannikainen, Peremptory Norms (jus cogens) in International Law:
Historical Development, Criteria, Present Status, Part III (1988).

Unlike the United Nations, in the GPA delegates would be elected by citizens rather than appointed by states. Because citizen elected representatives would not be controlled by national capitals, they would be unlikely to favor states maintaining their autonomy not to comply with international law. To the contrary, the institutional interests of GPA delegates would be in expanding that organization's powers. Over time they would, therefore, likely push for democratically approved international laws to be collectively binding on states and, even more importantly, on citizens directly. If citizens, loyal to an assembly elected by them, and that allowed for their lobbying and other participation, began to answer the call to directly follow democratically inspired international law, governments would lose the requisite influence over those they rely upon to defy international law.

C. War and the Prevention of Terrorism

In the age of weapons of mass destruction, the most dangerous deficiency of the global system is its propensity for political violence. A global parliament would provide a democratic substitute to achieving national security through domination and violence. In a global parliament there would be no unified states to counter, contain, or even attack other states. Rather, as occurs in other multinational parliaments such as in India, Canada or in the European Parliament delegates would likely break national ranks to vote along lines of interest and ideology.[12] Thus, fluid transnational parliamentary coalitions could begin to supplant conflict, including armed conflict, among states. If parliamentary decision-making proved itself successful, it is possible to imagine over time a genuine lessening of global tensions, and perhaps, if citizens gradually gained confidence in global democratic processes, meaningful disarmament.

Likewise, the GPA would offer disaffected citizens an alternative to terrorism and other forms of political violence. Those impassioned about perceived injustices would be less likely to feel forced to choose between surrender and resort to desperate tactics. Citizens would be able to stand for office, champion candidates and form coalitions to lobby the parliament. Those with diverse or opposing views would be brought into a give-and-take setting that would improve the chances for compromise and reconciliation. And when compromise was not possible, even those whose views did not prevail would more likely accept defeat out of a belief in the fairness of the

---

[12] For an examination of transnational voting in the European Parliament, see generally, Amie Kreppel, THE EUROPEAN PARLIAMENT AND SUPRANATIONAL PARTY SYSTEM: A STUDY IN INSTITUTIONAL DEVELOPMENT (2002).

process, and a knowledge that they could continue to press their cause on future occasions.

In particular, a Global Parliamentary Assembly would directly counter the attraction of antidemocratic extremists such as Al Qaeda. One important feature of liberal parliamentary process is its capacity to assimilate even those who do not share a pre-existing commitment to democracy. Because parliaments invite participation and have the ability to confer popular legitimacy on a policy position, experience suggests that even those with extreme agendas will often be drawn into the parliamentary process. Of course, the Osama bin Ladens of the planet will never accept the legitimacy of a global parliament. But their ability to attract a significant following would be diminished by the vitality of such an institution.

IV. Conclusion

In making today's case for a GPA, I have argued that such an institution is both feasible and highly desirable. Of course, the constraints of a conference where global democracy is only one among many topics has required that my presentation be at the most general level of abstraction. Many more questions at the threshold between feasibility and desirability must be satisfactorily answered for such a project to proceed to fruition.

For example, what voting method or methods for selecting parliamentarians should be adopted, and in this connection how would voting districts be determined? Likewise, how would a GPA ensure that the elections to select its membership are both free and fair? What rules of campaign finance would be appropriate and workable for a globally elected body?

In addition, how would the parliament be organized internally? What would be the role of political parties, and what systems would have to be put in place to ensure that the parliament's business was appropriately translated so both parliamentarians and citizens from all over the world could fully participate in the organization's business?

As important as it is that these matters be appropriately considered by scholars and those who wish to participate in building a GPA project, they are of the nature of the questions that all democratic polities must face and should not obscure the twin core realities that it has been my purpose to convey today: The global system is deficient in that it is undemocratic, ineffective, and prone toward political violence and a global democratic vision, practically and incrementally pursued, could make an important contribution toward overcoming these deficiencies.

2
On the Establishment and the Incremental Development of a GPA

On the Creation of a Global Peoples Assembly: Legitimacy and the Power of Popular Sovereignty

by Richard Falk and Andrew Strauss

Stanford Journal of International Law, 2000[*]

I. Introduction: The Vision of a Global Peoples Assembly

As recent street protests in Seattle and Washington, D.C.. have made increasingly clear, those who are active in global civil society are committed to the promotion of global democracy.[1] Many have undoubtedly at some point hoped that a worldwide popularly elected legislative assembly would be established. Until recently, such a prospect reeked of utopianism. The powers that be in the world — including those leaders who champion democracy in state/society relations — seemed clearly unreceptive to such an innovation. At this historical juncture we believe that the time for the establishment of a global assembly is ripening. We believe that our circumstances and values are raising a crucial new question: If democracy is so appropriate in the nation-state setting, why should not democratic procedures and institutions be extended to the global setting?

The road to this juncture has not been an easy one. Indeed, for more than a century the state system, as initially established and formalized by the Peace of Westphalia (1648), was dominated by the absolutist state usually headed by a royal monarch.[2] Only with the French Revolution was the democratic idea launched as the foundation of legitimate government. Even the American Revolution — with its fear of the tyranny of the majority and its limitations on citizenship and voting rights — was at first only partially committed to the democratic idea. But over time, and especially during the

[1] For further discussion of the Seattle and Washington protests, see *infra* text accompanying note 91.

[2] See RICHARD FALK, LAW IN AN EMERGING GLOBAL VILLAGE: A POST-WESTPHALIAN PERSPECTIVE 4 (1998).

last half of the twentieth century, the conviction grew that only a government that rests on the genuine consent of its people, as expressed through periodic, multi-party, fair elections, is legitimate. With the fall of the Soviet Union and the end of the Cold War, the idea that the only legitimate form of government is a democracy has steadily gained ground, leaving the conspicuous lack of global democratic institutions as the world's greatest political anomaly.[3]

That we suggest democracy should be extended to the international arena should not be taken as an unqualified endorsement of the quality of democratic governance in state/society settings.[4] Of course, not all states are democratically constituted in such a way as even to maintain the pretense of vesting ultimate authority in their citizenry. What is more, as in many countries where the real power lies with the military rather than with elected representatives of the people, authoritarian structures often persist behind the facade of constitutionalism.[5] In other countries, like the United States, monied interests vastly distort the representative process, as do national security doctrine and practice, which lend credence to broad claims of secrecy and even public deception.[6] Also, it is arguable that the discipline of global capital — the dynamics of economic globalization — is constraining democratic governance, and is giving rise to an era of "choiceless democracy."[7] Beyond all this,

---

[3] For influential depictions of the democratic idea as it has evolved in the West, see generally ROBERT DAHL, ON DEMOCRACY (1998); DAVID HELD, DEMOCRACY AND GLOBAL ORDER: FROM THE MODERN STATE TO COSMOPOLITAN GOVERNANCE (1995).

[4] *See The Global Democratic Prospect,* in THE GLOBAL RESURGENCE OF DEMOCRACY 247-324 (Larry Diamond & Marc Plattner eds., 2d ed. 1996) (containing essays discussing the problems plaguing the development of democratic systems of governance); *see also* RONALD DWORKIN, FREEDOM'S LAW 1-38 (1996) (examining the problem of protecting individual rights in majoritarian systems); Robert Lipkin, *Religious Justification in the American Communitarian Republic,* 25 CAP. U.L. REV. 765, 783-87 (1996) (discussing the need for communitarian values to improve the political discourse in democratic societies).

[5] *See* Fareed Zakaria, *The Rise of Illiberal Democracy,* FOREIGN AFF., Nov./Dec. 1997, at 22 (describing the emergence of a new breed of regime that governs in an authoritarian manner despite electoral validation). For a discussion of how the formalities of democracies can serve to disguise the actualities of military rule, see generally Richard Falk, *Democratic Disguise: Post-Cold War Authoritarianism,* in ALTERED STATES: A READER IN THE NEW WORLD ORDER 17 (Phyllis Bennis & Michel Moushabeck eds., 1993).

[6] For an authoritative and detailed account of the corrosive influence of money on U.S. politics, see CHARLES LEWIS & THE CENTER FOR PUBLIC INTEGRITY, THE BUYING OF THE PRESIDENT 2000 (2000); on the impact of national security bureaucracy, see RICHARD J. BARNET, ROOTS OF WAR (1972).

[7] For elaboration, see FALK, *supra* note 2, at 169; and RICHARD FALK, ON HUMANE GOVERNANCE: TOWARD A NEW GLOBAL POLITICS 104-33 (1995) [hereinafter FALK, ON HUMANE GOVERNANCE]. *See also generally* David Held, Democracy and Globalization, in REIMAGINING POLITICAL COMMUNITY 11 (Daniele Archibugi et al. eds., 1998).

there are questions about the dangers of democratic governance if the under-lying political culture is illiberal or militarist.[8]

Despite these reservations, we join in support of the democratic idea of governance, seeking to vitalize it at all levels of social interaction, and, in particular, to extend democracy globally.[9] We believe that an internationally elected global assembly could eventually overcome several of the fundamen-tal constraints that currently hinder the development of humane and effective global governance. The existence and empowerment of a Global Peoples Assembly (GPA) would, at the most general level, challenge the traditional claim of states that each has a sovereign right to act autonomously, regard-less of adverse external consequences. This challenge, ensuing straight from the most fundamental democratic precept that government derives its just powers from the consent of the governed, would undermine the claim that states are bound only by state-created international law, and then only when they give their consent.[10]

This means that state claims of a right to opt out of collective efforts to preserve the global commons, reduce or eliminate weapons systems, safe-guard human rights, or otherwise protect the global community could be authoritatively questioned. Not only might such an assembly lead over time to the establishment of beneficial and effective community law, but also its

---

[8] See FALK, *supra* note 2, at 165-66; FALK, ON HUMANE GOVERNANCE, *supra* note 7, at 70-74, 163. If the political culture of a country is militaristic, it sends "democratic" signals to leaders that encourage recourse to force. This is particularly serious for a country such as the United States, which occupies the role of global leader. For a general discussion of the rela-tionship between culture and democracy, see J.M. Balkin, *The Declaration and the Promise of a Democratic Culture*, 4 WID. L. SYMP. J. 167, 173 (1999) ("Democracy inheres not only in procedural mechanisms like universal suffrage but in cultural modes like dress, language, manners, and behavior. Political egalitarianism must be nourished by cultural egalitarian-ism.").

[9] There is an important challenge to democracy, at the interface of the state and the glob-al order, relating to the relevance of international law to the foreign policy process. The idea of the rule of law is integral to the realization of constitutional democracy internal to the state. In countries like the United States, however, evasions of the rule of law in external relations have been upheld by courts as constitutionally permissible. The political question doctrine, for example, has been validated by reference to the need for unity under the authority of the Pres-ident. As a result, in the United States legal scrutiny of the President's actions in foreign poli-cy has rarely taken place aside from "the court of public opinion," as in the latter stages of the Vietnam War. *See generally* Richard Falk, *The Extension of Law to Foreign Policy: The Next Constitutional Challenge*, in CONSTITUTIONALISM: THE PHILOSOPHICAL DIMENSION 205 (Alan Rosenbaum ed., 1988).

[10] *See* Louis Henkin, *International Law: Politics, Values and Functions*, 216 RECUEIL DES COURS D' ACADEMIE DE DROIT INT'L [COLLECTED COURSES OF THE HAGUE ACAD. OF INT'L L.] 27 (1989) ("[A] State is not subject to any external authority unless it has voluntarily consented to such authority.").

supranational lawmaking mode would itself be transformative. To the extent that citizen-elected representatives from different countries and civilizations convene formally in a climate of civility to advance mutual interests and address differences, peaceful resolution of conflict would tend to become institutionalized. Interest groups attempting to influence the GPA would quickly coalesce across national lines, eroding the strength of arbitrary and often dangerous national identities.[11] The normal parliamentary process of delegates working to build social consensus on issues would encourage the promotion of widely shared values over more parochial concerns and beliefs. Ultimately, such an assembly could lead the way to a global parliamentary system where social, political, and even cultural differences might come to be settled in as peaceful and fair a way as has been the experience within some of the world's more successful democratic societies.[12]

Alas, despite these advantages that should be clear even to those acting purely from the perspective of enlightened self-interest, most observers still are likely to dismiss the proposal for such an assembly as utopian. From a U.S. setting one can ask, "How, in a world where U.S. Senator Jesse Helms has a veto on the sole superpower's adherence to new international institutions and agreements, could such a step beyond the confines of sovereignty ever come about?" Even putting Senator Helms aside as a political anomaly, formidable resistance to any global innovation that seemed to subordinate sovereign rights to some higher external authority would remain in the United States and in many other countries as well.

One response to such resistance, easily obscured by our state-centric presuppositions, is that a GPA need not be established by a traditional interstate treaty arrangement. Globalization has generated an emergent global civil society composed of transnational business, labor, media, religious, and issue-oriented citizen advocacy networks with an expanding independent ca-

---

[11] Within the European Union's only popularly elected governing institution, the European Parliament, transnational coalitions among pressure groups as well as among parliamentarians have become common. *See Europe's Political Parties: The Slow March to Greater Bonding*, THE ECONOMIST, Mar. 6, 1999, at 49 (citing European Parliament voting statistics demonstrating that parliamentarians from pan-European political parties - such as the Christian Democrats and the Socialists - are increasingly voting along party lines regardless of their countries of origin, and that parliamentarians often vote along ideological rather than national lines).

[12] For further elaboration on how the existence of a GPA might help overcome the dysfunctions of the present international legal system, see generally Andrew Strauss, *Overcoming the Dysfunction of the Bifurcated Global System: The Promise of a Peoples Assembly*, 9 TRANSNAT'L. L. & CONTEMP. PROBS. 489 (1999) (discussing how a GPA could help improve the creation of, compliance with, and enforcement of international law). *See also* Richard Falk & Andrew Strauss, *Globalization Needs a Dose of Democracy*, INT'L HERALD TRIB., Oct. 5, 1999, at 8.

pacity to initiate and validate a GPA.[13] In one of the most significant, if not yet fully appreciated, developments of the post-Cold War era, global civil society operating in collaboration with certain like-minded states-has become a formidable political presence in international life, pushing forward several key progressive initiatives in the international arena.[14] While civil society has never undertaken a project of such magnitude as that of organizing a GPA, its accomplishments over the last decade have been impressive.[15]

In what follows, we wish to make the case that civil society is now capable of founding the Global Peoples Assembly, and that, because of its basis in popular legitimacy, the GPA would have the potential to play a major role in global governance.[16] A proposal of this scope of course raises other major

---

[13] The concept of civil society is ambiguous and has brought about some amount of confusion. For the purposes of this Article, we accept Larry Diamond's definition:

[Civil Society] is distinct from "society" in general in that it involves citizens *acting collectively in a public sphere* to express their interests, passions, and ideas, exchange information, achieve mutual goals, make demands on the state, and hold state officials accountable. Civil society is an intermediary entity, standing between the private sphere and the state. Thus, it excludes individual and family life, inward looking group activity (e.g. recreation, entertainment, or spirituality), the for-profit-making enterprise of individual business firms, and political efforts to take control of the state.

Larry Diamond, *Toward Democratic Consolidation*, in THE GLOBAL RESURGENCE OF DEMOCRACY 227, 228 (Larry Diamond & Marc Platter eds., 2d ed. 1996). The term "civil society organization" refers to one of the many individual organizations that operate within civil society and is often used interchangeably with the term "nongovernmental organization" (NGO). The two terms will be used synonymously in this Article.

[14] Of course, civil society's attempts to influence international governance did not begin with the end of the Cold War. For an historical examination of civil society's international initiatives, see generally Steve Charnovitz, *Two Centuries of Participation: NGOs and International Governance*, 18 MICH J. INT'L L. 183 (1997). Those states that have joined with civil society to promote the joint projects we refer to in Part II.A have been referred to as "like-minded." In this Article, we also apply the term "like-minded" to those states that might be recruited to help promote a GPA.

[15] Several works of recent vintage give a more general overview of this development than does our selective discussion in Part II.A. For some of the most informative and interesting commentary, see generally Peter J. Spiro, New Global Potentates: Nongovernmental Organizations and the "Unregulated" Marketplace, 18 CARDOZO. L. REV. 957 (1996); Peter J. Spiro, *New Global Communities: Nongovernmental Organizations in International Decision-Making Institutions*, WASH. Q., Winter 1995, at 45; Jessica T. Mathews, *Powershifts*, FOREIGN AFF., Jan./Feb. 1997, at 50; P. J. Simmons, *Learning to Live With NGOs*, 112 FOREIGN POL'Y 82 (Fall 1998); and ABRAM CHAYES & ANTONIA HANDLER CHAYES, THE NEW SOVEREIGNTY: COMPLIANCE WITH INTERNATIONAL REGULATORY AGREEMENTS 250-70 (1995).

[16] In fact, in a recent development, civil society has begun to come together to promote global democracy initiatives. While its nascent efforts are still highly protean and the emerging visions are still inchoate, something significant appears to be happening. Three initiatives we believe merit further attention. From Perugia, Italy, an organization called the Assembly of the United Nations of Peoples has attempted to bring civil society organizations together into

questions: How would civil society come together to organize, overcome opposition, and bring the assembly into existence? How would it function once in existence? What effect would it have on the structure of the global political order? This Article deals with these questions only in a rough, schematic fashion. Our intention here is to take the first step in conceptualizing a new possibility, a possibility that — if it is to be seriously explored — will require tremendous future thought and development, not only by ourselves, but by many different actors from all parts of the world.[17]

II. Realizing the Vision

A. The Accomplishments of Global Civil Society

Three major achievements of civil society during the 1990s stand out as illustrations of what is sometimes called "the new diplomacy," or "the new internationalism," although in each instance the gains are provisional and

a quasi-representative assembly. In the fall of 1999, with civil society organizations from 100 countries in attendance, it had its third assembly. Also noteworthy is the Global Peoples Assembly Movement that was launched at civil society's highly successful Hague Appeal for Peace in the spring of 1999. This organization had its first major assembly in Samoa in April, 2000, and in the midst of a great show of organizational energy has been enjoying rapid growth in membership. Like the Perugia initiative, its purpose is to prefigure a globally democratic institutional structure that would enable the peoples of the world to have a meaningful and effective voice in global governance. Also of importance, the highly regarded civil society organization, EarthAction, is organizing an NGO coalition tentatively called, "Citizens Century: Campaign for a Democratic U.N.," which as its name implies is looking to democratize global governance. Finally of importance, and linking these three civil society initiatives, was the Millennium NGO Forum. At the invitation of the United Nations Secretary-General, representatives of hundreds of civil society organizations convened this last May, 2000 at the United Nations Headquarters in New York. One of the primary stated goals of the forum was "to create an organizational structure whereby peoples of the world can participate effectively in global decision-making." The Forum's outcome will be reported on in September, 2000 by the Secretary-General to a special millennial assembly of states examining the future architecture of the global system of governance.

[17] In our writing we are, of course, influenced by our Western, primarily U.S., experience. With this knowledge in mind we very much invite discussion and collaboration with scholars and activists from all over the world. In particular, we invite such engagement from those in societies, most different from our own, outside of the West, who have historically been marginalized in the planning for new international institutions. It is our great hope that the force of the idea of global democracy will be such that citizens from all over the world will come to bring their personal and civilizational perspectives to a vigorous discussion about how it might best be achieved.

important qualifications are in order.[18] The first is the role global civil society played in establishing the global climate change framework convention. The second is its importance in bringing into force the convention outlawing antipersonnel landmines, and the third is the leadership it exercised in bringing about an agreement to establish an international criminal court.[19]

1. The Climate Change Convention

The 1992 U.N. Framework Convention on Climate Change (Climate Change Convention, or Convention) was adopted at the U.N. Conference on Environment and Development in Rio de Janeiro (Rio Conference).[20] The Climate Change Convention established nonbinding targets for reductions in greenhouse gases. The Convention's follow-up Kyoto Protocol, adopted in Kyoto, Japan in 1997, attempts to establish binding limitations on green-

---

[18] Civil society plays a central role in this new diplomacy, as distinguished from the old (exclusively state-centric) diplomacy. This new brand of diplomacy is characterized by a collaborative relationship between civil society and states dedicated to similar goals.

[19] Other less mammoth initiatives for which global civil society has been a driving force are too numerous to mention. For a general discussion of its evolving role in the areas of human rights, environmental law, and violence against women, see MARGARET E. KECK & KATHRYN SIKKINK, ACTIVISTS BEYOND BORDERS: ADVOCACY NETWORKS IN INTERNATIONAL POLITICS (1998). Transnational social forces played a significant part in encouraging the International Court of Justice to rule on the legality of nuclear weapons. *See* FALK, *supra* note 2, at 173-85.

[20] The U.N. Framework Convention on Climate Change was opened for signature at the U.N. Conference on Environment and Development in Rio de Janeiro, June 1992, 31 I.L.M. 814, 816 (1992) (entered into force March 1994).

More than 1400 NGOs were accredited to the Rio Conference. Some 25,000 other individuals attended the parallel Global NGO Forum. In addition to the Climate Change Convention, three other major international instruments were also concluded at the Rio Conference. The first was the Rio Declaration, a forward-looking, aspirational statement of principles for achieving an environmentally sustainable future. *See generally* David Wirth, *The Rio Declaration on Environment and Development: Two Steps Forward and One Back, or Vice Versa?*, 29 GA. L. REV. 599 (1995). The second was the Biodiversity Convention, a multilateral treaty whose purpose is to help protect the diversity of life forms on the planet from the harmful effects of human activity. *See generally* FIONA MCCONNELL, THE BIODIVERSITY CONVENTION: A NEGOTIATING HISTORY (1996). The third was AGENDA 21, a comprehensive work plan for achieving sustainable development. See AGENDA 21: THE EARTH SUMMIT STRATEGY TO SAVE OUR PLANET (Daniel Sitarz ed., 1993). For further discussion of Rio and the connection between climate change and sustainable development, see generally David Hodas, *The Climate Change Convention and Evolving Legal Models of Sustainable Development*, 13 PACE ENV'TL. L. REV. 75 (1995). For a comprehensive exploration of the climate change problem by the chairman of the Inter-Governmental Conference on Global Warming, see generally SIR JOHN HOUGHTON, GLOBAL WARMING: THE COMPLETE BRIEFING (2d ed. 1997).

house gases.[21] It has thus far received only a few of the ratifications necessary for it to enter into force.[22] Although the commitment of many countries, including the United States, to deal seriously with the problem remains in question, civil society — specifically environmental nongovernmental organizations (NGOs) has been indispensable to the progress that has thus far been made. The precedent for significant NGO involvement in the negotiation of environmental treaties was set in the very successful 1980s treaty negotiations over stratospheric ozone.[23] That success gave civil society the confidence to take on the politically more challenging problem of climate change.[24] As a first step, organizations such as the World Resources Institute and the International Union for Conservation of Nature laid the groundwork for climate change treaty discussions by supporting the conduct and dissemination of scientific studies. These studies became instrumental in helping to

[21] Kyoto Protocol to the United Nations Framework Convention on Climate Change, FCCC/CP/7/Add. 1, *reprinted at* 37 I.L.M. 2243 (1998).

[22] The Kyoto Protocol needs 55 parties representing at least 55% of global carbon emissions to come into force. To date only a handful of countries have ratified it. *See Kyoto Protocol Status of Ratification (as of 16 July 1999), available at* <http://www.unfccc.de/resource/-kpstats.pdf> (visited Aug. 13, 1999). Of particular importance, the U.S. Senate has not ratified the convention and there appears to be little short-term prospect of its doing so. *See Remember Global Warming?*, N.Y. TIMES, Nov. 11, 1998, at A26 (discussing prospects for Senate ratification).

[23] *See* Peter M. Haas, *Stratospheric Ozone: Regime Formation in Stages, in* POLAR POLITICS: CREATING INTERNATIONAL ENVIRONMENTAL REGIMES 152, 176 (Oran R. Young & Gail Osherenko eds., 1993) (describing how "a small, transnational group of scientists and policy makers" were instrumental in heightening public awareness and concern about ozone depletion, leading in turn to pressure on their governments to take further action). The influence of environmental civil society organizations on global environmental policy in the 1990s has gone far beyond affecting treaty negotiations. For a very good general discussion of the increasing importance of these organizations, see PAUL WAPNER, ENVIRONMENTAL ACTIVISM AND WORLD CIVIC POLITICS (1996); THE STATE AND SOCIAL POWER IN GLOBAL ENVIRONMENTAL POLITICS (Ronnie D. Lipschutz & Ken Conca eds., 1993).

[24] It was clear from the outset that even a modestly effective global warming treaty would have a significant impact on powerful industries. As predicted, many industrial players have been aggressively lobbying against global efforts to create binding limitations on greenhouse gases. *See, e.g., Subcommittee on National Economic Growth, Natural Resources and Regulatory Affairs of the House Government Reform and Oversight Committee*, Sept. 15, 1998 (testimony of William O'Keefe, Executive Vice President and Chief Operating Officer of the American Petroleum Institute). Opposition, however, has not been uniform. European industrial concerns have on the whole been more receptive to efforts to reduce global warming than have their U.S. counterparts and, within the United States, opposition even within affected industries has been varied.

establish a consensus on the serious danger of global warming.[25] Of particular importance, in 1985, the civil society organization International Council
of Scientific Unions, in cooperation with the U.N. Environment Programme
and the World Meteorological Organization, sponsored an international scientific climate change conference in Villach, Austria.[26] This conference and
its follow-up workshops played a major role in convincing the scientific
community that global warming was a real and urgent problem. The conference was also instrumental in establishing the Intergovernmental Panel on
Climate Change, which peer reviewed many of the later scientific and policy
studies upon which the drafters of the Climate Change Convention relied.[27]

Promoting the science that established the climate change problem was,
however, only the beginning. Largely as a result of the persistent advocacy
of global civil society, states agreed to place negotiations for a global climate
change treaty on the international agenda for action.[28] This advocacy began
in earnest in Toronto in 1988 at the Canadian-sponsored World Conference
on the Changing Atmosphere.[29] With significant NGO involvement, 300
scientists and policymakers from numerous countries, as well as the United
Nations and affiliated international organizations, mobilized the international
community by proposing specific limitations on global emissions of greenhouse gases.

With a clear vision of the problem and a proposed solution, the way was
set for the start of treaty negotiations. In 1990, the U.N. General Assembly

---

[25] *See* David Tolbert, *Global Climate Change and the Role of International Non-
Governmental Organisations*, in INTERNATIONAL LAW AND GLOBAL CLIMATE CHANGE 95, 98-
101 (Robin Churchill & David Freestone eds., 1991).

[26] *See generally* Report on the Villach Conference of the World Climate Impact Studies
Programme (1985), *summarized in* United Nations Environment Programme, 1985 Annual
Report of the Executive Director, U.N. Doc. UNEP/GC. 1412, at 70-71 (1986).

[27] *See* Jack Fitzgerald, *The Intergovernmental Panel on Climate Change: Taking the
First Steps Towards a Global Response*, 14 S. ILL. U. L.J. 231, 233 (1990); Tolbert, *supra* note
25, at 99.

[28] It is important to note that not all civil society organizations have championed a global
climate change regime. Some business-oriented groups have been especially opposed to the
coming into force of the Kyoto Protocol. The Cato Institute and the U.S. Chamber of Commerce, for example, have been high-profile opponents. *See Subcommittee on International
Economic Policy, Export and Trade Promotion of the Senate Foreign Relations Committee*,
June 26, 1997 (testimony of Patrick J. Michaels, Senior Fellow in Environmental Studies at
Cato Institute); U.S. Chamber of Commerce, *Clinton Climate Treaty A 'Lose-Lose' All Pain
and No Gain*, Dec. 11, 1997, *available at* <http://www.uschamber.com/policy/climatechange-
bulletin5.htm> (visited Aug. 20, 1999).

[29] Conference Statement, The Changing Atmosphere: Implications for Global Security,
WMO/OMM-No.710 (June 30, 1988), *reprinted in* INTERNATIONAL LAW AND GLOBAL CLI-
MATE CHANGE, *supra* note 25, at 367.

formally established the Intergovernmental Negotiating Committee for a Framework Convention on Global Warming.[30] Civil society organizations had a major effect on the negotiations by winning for their representatives the right to participate directly in the negotiations as "observers."[31] Not only did their observer status give civil society organizations direct access to the proceedings, but also it helped legitimize their ability to influence the negotiations. This influence occurred in a variety of ways. For example, in addition to traditional lobbying, environmentalists placed themselves at the heart of the Rio negotiating process by publishing a daily newspaper, summarizing the day's formal negotiating sessions, that governments relied on both to stay informed of developments in the negotiations and to float new proposals for advancing discussions.[32] Perhaps most consequentially, people associated with environmental organizations were given important policymaking positions inside government delegations, making it difficult to distinguish where the role of traditional sovereign players ended and that of civil society began.[33] The role that the Centre for International Environmental Law (CIEL) performed in advising small island states was particularly notable.[34] Due to their special vulnerability to rising sea levels (a predicted consequence of global warming), these countries had a particular interest in the negotiations, but very little in the way of in-house expertise, outside resources, and political clout. The Ford Foundation decided to help by making a grant to CIEL.[35] As a result, not only was CIEL able to assist these island states in coordinating positions and strategies, but also some of its lawyers actually were able to negotiate on behalf of certain countries.[36] Thus, the organization was able to exert tremendous influence on the overall negotiations.[37]

---

[30] *See* G.A. Res. 45/212, U.N. EC.C.C. (Dec. 21, 1990).

[31] *See* Ann Doherty, *The Role of Nongovernmental Organizations in UNCED*, *in* NEGO- TIATING INTERNATIONAL REGIMES 199, 203-08 (Bertram 1. Spector et al. eds., 1994) (discussing NGO accreditation to and participation in the Rio Conference).

[32] *See generally* ECO, *available at* Climate Action Network. <http://www.igc.org/-climate/Eco.html> (visited May 30, 2000).

[33] The U.S. delegation to the Rio Conference itself included more than a dozen such members.

[34] CIEL was founded in 1989 to advocate for the protection of the global environment. The Foundation for International Environmental Law and Development subsequently split off from CIEL. Both organizations continue to exert an important influence on international environmental policy.

[35] *See* CHAYES & CHAYES, *supra* note 15, at 260.

[36] CIEL, as well as representatives from other civil society organizations, also became directly involved in treaty drafting. *See id.* at 261. This was not the first time that environmental groups had played such a role. For example, Laura Kosloff and Mark Texler report that "NGOs, particularly the International Union for the Conservation of Nature (IUCN), were in

Civil society is now involved in the struggle to build the broad-based public support that is necessary to secure the ratifications required to bring into force the Kyoto Protocol, with its binding emissions limitations that were absent from the Climate Change Convention. It is far from assured, however, that crucial countries will ratify it, and that, even if extensively ratified, it will be consistently implemented.[38] It is, therefore, not yet possible to reach any definitive conclusions about the extent of the impact of the evolving climate change regime. Clearly, however, global civil society has played an impressive role in the progress that has thus far been made.[39]

2. The Convention Outlawing Anti-Personnel Land Mines

Global civil society was also a major catalyst for the recently entered-into-force convention outlawing anti-personnel landmines. Civil society's effort was notable because of the extent to which it influenced the treatymaking process and because of the speed with which it was able to bring participants successfully to conclude a treaty. This contribution was prominently acknowledged when the Nobel Foundation awarded the 1997 Nobel Peace Prize jointly to the International Campaign to Ban Land Mines (ICBL), a broad-based NGO coalition, and its coordinator, Jody Williams.[40]

Civil society's efforts began in October 1992 when six NGOs met on the heels of the Rio Conference to plan what would become the ICBL.[41] Almost

---

fact pivotal in the drafting of [the Convention on International Trade in Endangered Species]." Laura Kosloff & Mark Texler, Comment, *The Convention on International Trade in Endangered Species: No Carrot, But Where's the Stick?*, 17 ENVTL. L. REP. 10222 (1987).

[37] For further discussion of the role of CIEL in representing island states in the climate change negotiations, see CHAYES & CHAYES, *supra* note 15, at 260-62.

[38] Efforts to monitor state compliance with the nonbinding targets established in the Climate Change Convention are ongoing. Civil society organizations have been actively involved in such monitoring. *See* Climate Action Network, *Furthering the Objectives of the Framework Convention on Climate Change: The Role of Environmental NGOs* (Report prepared for the Subsidiary Body for Scientific and Technological Advice meeting in March of 1996); *see also* Juli Abouchar, *A Foot in the Door: The Role of Environmental Non-Governmental Organizations in Increasing Compliance with International Agreements*, 23 ALTERNATIVES 1. 20 (Mar. 1979) (discussing how failures of traditional international enforcement mechanisms have provided the basis for an increasing enforcement role for NGOs).

[39] For further comprehensive discussion on the role of NGOs in the climate change negotiations, *see generally* Chiara Giorgetti, *From Rio to Kyoto: A Study of the Involvement of Non-Governmental Organizations in the Negotiations on Climate Change*, 7 N.Y.U. ENVTL. L.J. 201 (1999).

[40] *See* Craig Turner & Norman Kempster, *Land Mine Crusader Wins Nobel Prize*, MIAMI HERALD, Oct 11, 1997, at A1.

[41] *See* Maxwell Cameron, *Democratization of Foreign Policy: The Ottawa Process as a Model, in* TO WALK WITHOUT FEAR 424, 431 (Maxwell Cameron et al. eds., 1998).

immediately, through a series of worldwide civil society conferences and seminars, the ICBL began a concerted effort to build public support for a global ban on landmines. As momentum built in late 1993, French members helped induce the French government to initiate a multilateral review of the landmines problem pursuant to the U.N. Convention on Prohibitions or Restrictions on the Use of Certain Conventional Weapons, the only existing treaty attempting to control the use of landmines.[42] The treaty was vague and permissive, having had no meaningful effect on the landmines problem, but the ICBL usefully seized upon the review process contained in the treaty as a way to focus further attention on the landmines challenge.[43] The review ended after two and one-half years without meaningful progress having been attained. Landmines had, however, been put high on the international agenda, and the ICBL was poised to make its major impact.

The organization initiated a public relations campaign to disseminate a well-defined humanitarian message, often highlighting its message with poignant images of the suffering caused by landmines.[44] ICBL eventually enlisted international personalities such as Princess Diana, Archbishop Desmond Tutu, and General Norman Schwarzkopf to champion the issue and provide media salience.[45]

Complementing this global effort was considerable grassroots organizational energy. A member of the ICBL, Mines Action Canada, was particularly active. It organized a massive Canadian petition and letter writing campaign that succeeded in encouraging the Canadian government to take a

[42] United Nations Convention on Prohibitions or Restrictions on the Use of Certain Conventional Weapons, opened for signature Apr. 10, 1981, U.N. Doc. A/Conf.95/15 (1980), *reprinted in* 19 I.L.M. 1523 (1980). See Michael Matheson, *The Revision of the Mines Protocol*, 91 AM. J. INT'L L. 158, 159 (1997) (discussing the French role in initiating the review of the convention).

[43] Protocol II of the United Nations Convention on Prohibitions or Restrictions on the Use of Certain Conventional Weapons included landmines among the list of weapons it proscribed. While Protocol II established certain broad limitations on the use of landmines, it did not limit the production, sale, or possession of these weapons. *See* Convention on Prohibitions or Restrictions on the Use of Certain Conventional Weapons, *supra* note 42, at 1529, 1530-31. For further discussion of Protocol 11, *see* R. J. Araujo, *Anti-Person Mines and Peremptory Norms of International Law: Argument and Catalyst*, 30 VAND. J. TRANSNAT'L L. 1, 20-22 (1997).

[44] These included images of prosthetic limbs lining hospital walls, piles of unworn shoes, and of child amputees speaking with Princess Diana. *See* Maxwell Cameron et al., To Walk Without Fear; in TO WALK WITHOUT FEAR, *supra* note 41.

[45] Civil society's public relations efforts were not limited to atmospherics. It gained the upper hand in the substantive public policy debate by effectively countering anti-treaty pronouncements by the United States. See Ken Roth, *New Minefields for N.G.O.'s*, THE NATION, Apr. 13, 1998, at 22 (describing how civil society organizations responded to various arguments made by the U.S. government against the treaty).

leadership role on the issue.[46] Canadian officials initiated meetings between NGOs and pro-ban states. These meetings supplemented direct lobbying of national governments by ICBL members. When such lobbying was successful and brought about favorable shifts in government policy in one country, the ICBL created momentum by immediately persuading other governments of the changing political climate.[47] These efforts ultimately culminated in a Canadian call for a meeting in Ottawa where pro-ban states could coalesce.[48] With seventy-four states attending, and the ICBL playing a major diplomatic role in the conference, the 1996 Ottawa Conference was a singular success.[49] It began a diplomatic process that ultimately led to Oslo where, within a short period of time, participants adopted an agreement to prohibit the use, production, development, sale, and stockpiling of landmines. Originally signed by 122 countries, the agreement achieved the requisite ratifications and, entered into force in March 1999, which is very quick for a controversial multilateral treaty.[50]

Here too, while taking comfort in the role of global civil society, we cannot report unqualified success. Major producers of landmines, including the United States and China, have not yet signed or ratified the treaty. The compliance machinery in the treaty is weak, leaving implementation on a largely voluntary basis. It is difficult, at this stage, fully to evaluate the achievement. It may be that the landmines campaign has created such a strong anti-landmines ethos as to make the treaty itself almost superfluous, or at least self-enforcing. But it may also be that its vague prohibitions will be put aside under battlefield pressures, or by governments hard-pressed to find low technology and inexpensive ways of engaging in warfare.

[46] For an account of the Canadian campaign by two of its leaders, see Valerie Wannington & Celina Tuttle, The Canadian Campaign, in TO WALK WITHOUT FEAR, *supra* note 41, at 48. The success of Mines Action Canada was very much aided by the fact that its calls fell on the ears of Lloyd Axworthy, Canadian Minister of Foreign Affairs. Axworthy was unusually dedicated to global reform and seized upon the landmines issue as an opportunity to establish a new Canadian role in the global policy community. *See id.* at 56.

[47] *See generally* Cameron, *supra* note 41, at 424.

[48] *See* Shawn Roberts, *No Exceptions, No Reservations, No Loopholes: The Campaign for the 1997 Convention on the Prohibition of the Development, Production, Stockpiling, Transfer and Use of Anti-Personnel Mines and on Their Destruction*, 9 COLO. J. INT'L ENVTL. L. & POL'Y 371, 379-80 (1998). For further discussion of Canada's role, see Robert Muller, *New Partnerships for a New World Order: NGOs, State Actors, and International Law in the Post-Cold War World*, 27 HOFSTRA L. REV. 21, 24-25 (1998).

[49] For further elaboration, see Bob Lawson, Toward a New Multilateralism: Canada and the Landmine Ban, BEHIND THE HEADLINES, Summer 1997, at 18, 20-21

[50] *See* Convention on the Prohibition of the Use, Stockpiling, Production and Transfer of Anti-Personnel Mines and on Their Destruction, Sept. 18, 1997, *available at* <http://www.un-.org/Depts/Landmine/UNDocs/ban_trty.html> (visited Mar. 30, 2000).

3. The Proposed Treaty to Create a
Permanent International Criminal Court

The achievement with the most far-reaching implications for the structure of the international order is also global civil society's most recent — the realization of an agreement for an international criminal court.[51] In the summer of 1998, 136 NGOs under the umbrella of the NGO, Coalition for an International Criminal Court, were accredited as observers by the U.N. conference in Rome that adopted the treaty.[52] Upon ratification by the requisite sixty countries, for the first time in history there will be a permanent independent international court capable of trying individuals responsible for genocide, crimes against humanity, and war crimes.[53]

The creation of a permanent international criminal court had been discussed on and off since the end of World War II.[54] The gathering momentum that led to Rome was only created, however, because public opinion led by civil society pressured U.N. Security Council members to establish ad-hoc tribunals in the wake of the horrors of Bosnia and Rwanda.[55]

A good part of civil society's success in promoting the international criminal court is due to the skilled way in which it made itself nearly indispensable to the negotiating process.[56] Because civil society's representatives to the

---

[51] For a description of the historical events leading up to the negotiations for the International Criminal Court, *see generally* ARYEH NEIER, WAR CRIMES: BRUTALITY, GENOCIDE, TERROR, AND THE STRUGGLE FOR JUSTICE (1998). Neier is President of George Soros's Open Society Institute, a civil society organization. His authorship of this book is thus in itself representative of the effort by global civil society to advance the cause of a permanent international criminal court.

[52] While 136 NGOs were officially accredited, 238 NGOs were actually represented in Rome. *See* M. Cherif Bassiouni, *Historical Survey: 1914-1998*, in STATUTE OF THE INTERNATIONAL CRIMINAL COURT: A DOCUMENTARY HISTORY 1, 26 n. 135 (M. Cherif Bassiouni ed. 1998).

[53] *See* Rome Statute for the International Criminal Court, U.N. Doc. A/Conf.183/9 (1998), *reprinted in* STATUTE OF THE INTERNATIONAL CRIMINAL COURT, *supra* note 52, at 37.

[54] The prominence of the Nuremberg and Tokyo war crimes tribunals at the end of World War II brought the possibility of a permanent tribunal into the realm of discussion. For an extensive accounting of the Nuremberg trials by a former prosecutor at Nuremberg and a distinguished jurist, see TELFORD TAYLOR, THE ANATOMY OF THE NUREMBERG TRIALS (1992).

[55] For a brief discussion of the political and conceptual relationship between the tribunals for the former Yugoslavia and Rwanda and the proposed International Criminal Court, see Roger S. Clark & Madeleine Sann, *Coping with the Ultimate Evil Through the Criminal Law*, 7 CRIM. L.F. 1 (1996).

[56] In this Article, we have chosen to focus on the key role that civil society played in fashioning the treaty negotiations that took place in Rome. For a more general background discussion of the historic role that civil society played in keeping the "Nuremberg idea" alive

Rome Conference included respected academic experts and former government policymakers, its representatives could address the many highly technical issues with great authority.[57] Many governments relied on these expert assessments of specific problems, thereby giving civil society a tremendous influence on framing the overall discussion.[58] Emulating the largely successful model from the Rio Conference, civil society made itself the information and communications center of the negotiations. It was integrally involved in publishing a conference newspaper that delegates used to float proposals and to stay abreast of developments in the negotiations.[59] Another opportunity for civil society to play a strategic communications role emerged because substantive negotiations took place in many different working groups. Smaller countries were not capable of sending representatives to all the groups and were unable to remain current with respect to these multifaceted negotiations.[60] Civil society responded by organizing itself into various teams, each responsible for monitoring a particular working group. These teams provided regular formal and informal briefings, and many governments came to rely on these briefings to shape their positions.[61]

As in Rio and Ottawa, representatives of civil society played important roles on governmental delegations. Perhaps their most direct influence in this regard came as a result of the John D. and Catherine T. MacArthur Foundation funding twenty-four delegates from twenty-two small developing countries. Cherif Bassiouni, one of the major U.S.-based academic supporters of the court, was involved in working directly with these countries.[62]

over the last fifty years, *see* Richard Falk, *Telford Taylor and the Legacy of Nuremberg*, 37 COLUM. J. TRANSNAT'L. L. 693, 716-21 (1999).

[57] *See* M. Cherif Bassiouni, *Negotiating the Treaty of Rome on the Establishment of an International Criminal Court*, 32 CORNELL INT'L. L .J. 443, 455 (1999) (identifying the major civil society organizations whose representatives played important roles).

[58] *See* John Washburn, *The Negotiation of the Rome Statute for the International Criminal Court and International Lawmaking in the 21st Century*, 11 PACE INT'L. L. REV. 361, 368-69 (1999) (explaining how NGOs enhanced the effectiveness of the like-minded states).

[59] *See* TERRAVIVA, available at Inter Press Service, <http://www.ips.org/icc/index.htm> (visited May 30,2000).

[60] *See* Bassiouni, *supra* note 52, at 29 (recounting the establishment of small, informal working groups at the Rome Conference and the organizational difficulties such groups presented to the smaller delegations).

[61] *See* Washburn, *supra* note 58, at 368-69.

[62] M. Cherif Bassiouni, a Professor at DePaul University College of Law, was also a member of the Egyptian delegation to the conference and headed up the drafting committee. See Henry T. King & Theodore C. Theofrastous, *From Nuremberg to Rome: A Step Backward for U.S. Foreign Policy*, 31 CASE W. RES. J. INT'L. L. 47, 96 97 n. 201 (1999)

We are indebted to Professor Michael Scharf, who formerly worked on war crimes issues in the Legal Advisor's Office of the U.S. State Department, for bringing to our attention

If civil society was helpful to governments by providing expertise, information, and resources, it also was potentially the greatest adversary of wavering governments. As several governments stung by harsh criticism during the negotiations learned, civil society had the capacity to mobilize public pressure against those who became recalcitrant. Civil society issued a daily Internet newsletter, transmitted to thousands of subscribers, that informed and coordinated the activities of constituents spread around the world who were poised to apply pressure directly in national capitals.[63] In addition, because the press was excluded from many of the more substantive sessions, representatives of civil society took advantage of the opportunity to become the media's primary source of unofficial information as well as a crucial check on government pronouncements.

It is true that the treaty that emerged from Rome contains concessions to geopolitical forces, including giving some control over its operation to the U.N. Security Council. Even with these compromises, however, the treaty is a momentous achievement. The expected existence of the international criminal court will finally establish that not only states but also individuals are responsible to the international community for gross violations of human rights. To make government leaders consistently accountable in this way is to modify the basic constitutional premise of a world order based on sovereign states. The struggle is, of course, far from over. States must now ratify, and then implement, the treaty. Although a big step was taken at Rome, there is a long way ahead before a functioning international criminal court comes into existence.

Without a doubt the accomplishments of global civil society indicate that it is coming of age and is capable of promoting significant global reform. In recognizing this new source of reformist influence we do not wish to essentialize or romanticize global civil society. It would be a serious misconception of the diversity of outlook in global civil society to attribute to it a single, enlightened viewpoint, as militant encounters between "right to life" and "pro-choice" groups have illustrated with high drama in the struggle over U.S. public policy on abortion. There are not only opposing orientations on a variety of global issues, but also disagreement as to tactics and substance among those who are committed to similar values. For instance, some transnational activists are focused exclusively on a given category of weaponry, say, nuclear weapons or biological weapons, while others emphasize disarmament or even the elimination of war as a social institution. For the

---

the role played by the John D. and Catherine T. MacArthur Foundation and Professor Bassiouni.

[63] See generally ON THE RECORD, available at The Advocacy Project, <http://www.advocacynet.org/cgi-bin/browse.pl?id=otr> (visited May 30, 2000).

Global Peoples Assembly to be successfully brought into existence, however, neither unity of perspective nor even the uncommon virtue sometimes misleadingly attributed to civil society is necessary.[64] To bring the project to fruition, it is only necessary that enough of the diverse elements of civil society effectively unite around a common vision of a GPA.

B. Securing the Support of like-Minded States:
The Encouraging Reality of the European Parliament

Despite the political accomplishments of civil society, its financial and logistical resources still pale in comparison to those at the disposal of states. Therefore, even if civil society were to unite in the way suggested, its efforts would be made far easier if some states could be recruited to help in the venture. It is encouraging that, in all of the civil society led successes of the 1990s that we have discussed, civil society was able to enlist important core support from at least some like-minded states. But is it realistic to think that sovereign states — no matter what their past record on certain projects – could be convinced to work with civil society to create an institution as ultimately threatening to the principle of state autonomy as is a Global Peoples Assembly? In fact, we believe that one important basis for treating this undertaking as plausible arises from the European experience with the establishment and evolution of an analogous institution of regional scope.

Three hundred and fifty million European citizens from the fifteen European Union counties are presently represented by a popularly elected assembly called the European Parliament (Parliament). The Parliament is not a product of the post-Cold War 1990s, but goes back to the earliest days of European integration following World War II.[65] When European integration began in earnest with the Treaty of Rome in 1957, the Parliament was given its place alongside the European Council and the European Commission as one of the three lawmaking bodies.[66] The Parliament, however, was not pop-

---

[64] The Nation recently featured a lively exchange between David Rieff and Michael Clough that brings into sharp relief much of the controversy surrounding the nature and implications of the ascent of global civil society. *See generally* David Rieff, *Civil Society and the Future of the Nation-State: The False Dawn of Civil Society*, THE NATION, Feb. 22, 1999, at 11; Michael Clough, *Civil Society and the Future of the Nation-State: Reflections on Civil Society*, THE NATION, Feb. 22, 1999, at 16. The debate precipitated responses from Kenneth Roth and Peter Weiss, two prominent representatives of civil society, as well as a rebuttal by Rieff. See Letters, THE NATION, March 29, 1999, at 2.

[65] *See generally* FOUNDATIONS OF DEMOCRACY IN THE EUROPEAN UNION: FROM THE GENESIS OF PARLIAMENTARY DEMOCRACY TO THE EUROPEAN PARLIAMENT (John Pinder ed., 1999) [hereinafter FOUNDATIONS OF DEMOCRACY].

[66] *See* Treaty Establishing the European Community (Treaty of Rome), Apr. 18, 1951, 261 U.N.T.S. 140 [hereinafter Treaty of Rome]. For an excellent reference on the overall

ularly elected, but rather delegates were appointed by national parliaments.[67] Moreover, the Parliament was politically quite weak and was not taken seriously as an element of European architecture. It came into being very much the poor stepchild of the other two governing institutions.

In 1979, citizens for the first time were given the right to elect their representatives to the Parliament.[68] Since that time, the Parliament has gradually gained expanded powers with successive European integration treaties. For years critics scoffed at the Parliament as a virtually meaningless body composed of incompetent and self-serving representatives with no meaningful influence. In light, however, of significant additional powers given to the body in the Maastricht Treaty and the Treaty of Amsterdam, the world is finally beginning to take notice of it.[69] Last year, in a watershed event, the whole of the European Commission resigned in response to pressure from

constitutional structure of the European Union, see generally THE CONSTITUTIONAL LAW OF THE EUROPEAN UNION (James D. Dinnage & John F. Murphy eds., 1996).

[67] *See* Richard Corbett, MEP, *The European Parliament and the Idea of European Representative Government,* in FOUNDATIONS OF DEMOCRACY, *supra* note 65, at 87, 90-93.

[68] Elections had actually been contemplated since the time of the Treaty of Rome, which provided that the "Assembly," as it was then called, would eventually be elected by universal suffrage. *See* Treaty of Rome, *supra* note 66, art. 138.

[69] Both the 1993 Maastricht Treaty and the later 1997 Treaty of Amsterdam significantly enhanced the formal legislative powers of the Parliament. Under the Maastricht Treaty, the Parliament was given so-called "co-decision" with the European Council (i.e., the power to amend or veto legislation in certain specified legislative areas such as education, cultural affairs, and public health). This was in addition to the right of "consultation" which the Parliament had previously gained in certain areas. Consultation, when it operates, requires that Parliament's opinion must be obtained before the European Council may adopt a legislative proposal from the Commission. Also, before Maastricht, Parliament had been able to participate in the European Union legislative process through a "co-operation procedure" which allowed it to reject the Council's draft legislation in other specified areas if its opinions were not sufficiently taken into account. (Cooperation currently only applies in certain areas of monetary union.) Predating Maastricht, Parliament had gained the power to force the resignation of the entire Commission. It has also historically had significant power over the European Union's budget. Treaty on European Union, Feb. 7, 1992, 1 C.M.L.R. 719, *reprinted in* 31 I.L.M. 247, 256 (1992) (amending Treaty Establishing the European Economic Community, Mar. 25, 1957,298 U.N.T.S. 11).

The Treaty of Amsterdam further expanded the powers of the Parliament. It more than doubled the substantive matters over which the Parliament has co-decision power, increasing its influence in such areas as employment policy, consumer protection, transportation, and the environment. The Parliament now has the power to approve the nominee for the European Commission presidency. It also gained the general power to approve international agreements between the European Union and third party states and the specific power to approve the accession of new member states to the European Union. See generally Treaty of Amsterdam amending the Treaty on European Union, the Treaties Establishing the European Communities and Certain Related Acts, Nov. 10, 1997, 1997 O.J. (C 340) 1, *available at* <http://europa.eu.int/> (visited May 30, 2000).

the Parliament over bribery and corruption charges.[70] There is a newfound sense that the Parliament really does matter and will do so increasingly.[71]

In last year's elections to the Parliament voter turnout was disappointingly low. Nevertheless, despite the fits and starts inherent in creating the first popularly elected transnational assembly, the European experiment clearly demonstrates that at least some states would potentially be receptive to participating in creating a GPA that could gradually assume real powers. True, Europe is far more homogeneous than the world, and its legislative efforts were complemented by ambitious movement toward impressive degrees of economic integration, but still the example of Europe is encouraging with respect to the creation of a legislative institution of global scope. Certainly the existence of the Parliament challenges the view that the creation of a transnational assembly akin to a GPA is an impossible utopian dream that would never enlist the support of sovereign states.

In fact, the same European Union states that have promoted the Parliament are among the most likely to lend their support to the creation of a GPA. Because of their citizens' experience with the European Union, a popular allegiance to absolute and supreme national sovereignty is becoming less and less of a living ideological tenet around which national identity is organized. Rather, this absolute allegiance is being replaced with a political culture that accepts supranational institutions as necessary to meeting world order challenges. For the citizenry of these European countries, the GPA would in many ways be a logical next step toward greater political harmony. We do not mean to suggest that gaining the support of countries in Europe or elsewhere would be easy, but only that — with the type of hard political work that went into civil society's achievements of the 1990s — it is possible and is worth the attempt.

Unlike all of the efforts of the 1990s, however, a traditional treaty-based approach is not presently a viable way to initiate a GPA. Because the near universal state support necessary to launch a worldwide assembly by treaty would most likely not be forthcoming, we argue that civil society, with the help of willing states, should itself create this assembly.[72] This novel ap-

---

[70] *See* Corbett, *supra* note 67, at 106.

[71] We discuss in Part IV.A infra why the European Parliament has been given an ever-expanding role despite its difficulties gaining credibility. We suggest that its empowerment has to a large extent been a response by European elites to resistance by the people of Europe to increasingly important European Union policy being set in an anti-democratic manner solely by "faceless bureaucrats" in Brussels.

[72] While it is unlikely that most states would at present be willing to adhere to a traditional treaty establishing a GPA, it may well be that the most effective strategy for bringing such an organization into existence would be by treaty created by those states enlisted to support the GPA. Such an approach might most effectively both overcome logistical and

proach to creating a supranational lawmaking institution presents a basic question: Even if civil society can overcome the financial and logistical difficulties of organizing a GPA, does not any institution staking a claim to transnational governance have to be validated by inter-state treaty? No one could imagine, for example, an international criminal court established by Amnesty International and other independent human rights organizations as having any claim to exercise binding authority in actual criminal cases.[73] Uniquely, a GPA would have a claim to authority independent of whether or not it received the formal blessings of the state system. This claim, we will now argue, rests on the degree to which popular sovereignty is becoming the foundation for governmental legitimacy in today's world, and can be extended to institution building on a global scale.

III. Legitimacy and the Global Peoples Assembly

Legitimacy helps explain why people obey.[74] In The Power of Legitimacy Among Nations, Professor Thomas Franck adapts what he regards as "a par-

---

organizational barriers to the assembly's creation and authoritatively establish its initial structure. The treaty could provide that as the assembly became known and gained acceptance, future countries could allow their citizens to vote by acceding to the treaty. Alternatively, modern notions of contemporary sovereignty, discussed infra Part III, allow for the founding states to provide that citizens from non-party states vote for representatives to the assembly. The GPA's gradual achievement of legislative authority under either treaty-based approach to enfranchising the global citizenry would arise from the dynamics of empowerment that we discuss infra Part IV.

[73] The Lelio Basso International Foundation for the Rights and Liberation of Peoples in Rome did, in fact, establish a "Permanent Peoples Tribunal" without state sanction in 1979. It engages in trial-like proceedings to highlight great travesties and injustices in the world. For example, it has ruled on the Armenian genocide, exploitative labor practices in developing countries, and the Union Carbide disaster in Bhopal, India. Although some of its judgments have received a good deal of publicity, the international community has never regarded them as binding. *See* Lelio Basso International Foundation for the Rights and Liberation of Peoples, <http://www.grisnet.it/filb/filbeng.html> (visited Feb. 18, 2000). For further discussion of its role in the development of the "law of humanity," *see generally* FALK, *supra* note 2, at 34, 43. For representative Permanent Peoples Tribunal cases, *see generally* PHILIPPINES: REPRESSION AND RESISTANCE (Marlene Dixon ed., 1981); ON TRIAL: REAGAN'S WAR AGAINST NICARAGUA (Marlene Dixon ed., 1985); GUATEMALA: TYRANNY ON TRIAL (Susanne Jonas et al., eds. 1984). For a consideration of jurisprudential foundations of a judicial process not constituted by governments, *see* ANTONIO CASSESE & EDMOND JOWE, POUR UN DROIT DES PEUPLES (1978). A comparable ad-hoc initiative on the legality of nuclear weapons was convened in London. *See* THE BOMB AND THE LAW (Geoffrey Darnton ed., 1989).

[74] For Professor Thomas Franck and others, legitimacy explains what, beyond a coercively enforced sovereign command, creates habitual compliance with law. *See*, e.g., THOMAS

tial definition of legitimacy to . . . the international system: a property of a rule or rule-making institution which itself exerts a pull towards compliance on those addressed normatively."[75] The voluntary compliance to which Professor Franck refers arises either because the actual rules themselves are in accord with what people believe, or because people accept as valid the source of the rulemaking institution's claim to the exercise of authority.[76] For example, the fifteenth century English citizenry might have complied with a royal decree criminalizing the practice of witchcraft either because the citizenry believed that witchcraft was evil or because it accepted as valid the claimed source of the crown's lawmaking authority — that is, that the monarch was appointed by God.

In contemplating the legitimacy of a GPA organized and constituted by grassroots action, the question becomes: What currently accepted source or sources of authority could confer upon a transnational organization of this type the authority to create international norms binding on governments? To date, the presumed answer has been state consent. Planners contemplating transnational organizations in today's world — where a deep-seated belief in state sovereignty remains pervasive — are accustomed to assuming that the authority of transnational organizations will be accepted as legitimate only to the extent that states have consented (usually by way of treaty) to create and

---

FRANCK, THE POWER OF LEGITIMACY AMONG NATIONS 16 (1990) (referring to this understanding of legitimacy as common to both Dworkin and Habermas).

One problem that permeates the use of the term "legitimacy" is discerning whether it is meant descriptively, in the way that Max Weber used the term, to denote when people are predisposed to obey rules without reference to whether the rules or rulemaking institutions are good or bad, or whether it is meant normatively. We are using the term in the former sociological sense to demonstrate that rules promulgated by the GPA would likely be obeyed. Obviously, the impetus for this Article comes from the fact that we are strongly supportive of such an assembly.

[75] FRANCK, *supra* note 74, at 16. Professor Franck elaborates:

Deployed by students of national legal systems, the concept of legitimacy is often used to postulate and explain what, other than a command and its enforcement, is required to create a propensity among citizens generally to obey the rulers and the rules. The internationalist ought to feel both comfortable with, and stimulated by, this notion of legitimacy as the noncoercive factor, or bundle of factors, predisposing toward voluntary obedience.

Id.

[76] This differentiation of reasons for voluntary compliance is most closely associated with the ideas of H.L.A. Hart, which are discussed and applied to the international order at some length in Franck's book. See FRANCK, supra note 74, at 183-94. In the words of Hart, the distinction we are making is between the "rules of recognition," those primary rules that establish the rulemaking institutions, and the rules that those institutions create. *See* H.L.A. HART, THE CONCEPT OF LAW 97-114 (1961).

be bound by such organizations.[77] While state consent has no doubt served to confer some measure of legitimacy upon transnational institutions, the belief that state consent is the exclusive means of legitimizing transnational institutions is anachronistic. It assumes a belief in what we call "classic sovereignty," an assumption that is in fact at odds with what most people today actually believe and with numerous areas of international practice.

Classic sovereignty was an outgrowth of the early modern belief that the king was the exclusive source of all temporal law.[78] This was not a democratic belief. The king, as sovereign, was accepted as the personification of the state who ruled over his subjects below and could be subject to no higher secular authority without his consent.[79] Those believing in classic sovereignty would naturally consider the sovereign's consent as necessary to legitimize transnational organizations and would reject the notion that subjects of the various states could create transnational organizations by their own initiative.

Today, however, there is a growing acceptance of what we call "contemporary sovereignty," which reconciles the state system with the modern commitment to democracy and human rights.[80] Modern democratic beliefs

---

[77] For further observations on the continuing and pervasive emotional allegiance to the sovereign state, see FALK, ON HUMANE GOVERNANCE, *supra* note 7, at 79-103.

[78] For one of the classic works on the development of this idea, *see* JOHN N. FIGGIS, THE DIVINE RIGHT OF KINGS (Peter Smith Publisher, Inc. 1970) (1896).

[79] This contrasts with the premodern medieval idea that the Holy Roman Church possessed the highest normative authority. See FALK, supra note 2, at 26. For what is usually regarded as the seminal work of the sixteenth century evidencing this understanding of sovereignty, *see* JEAN BODIN, SIX BOOKS OF THE COMMONWEALTH (M .J. Tooley trans., 1955) (1576). By the nineteenth century, the all-powerful sovereign was no longer necessarily personified in a monarch, but could be represented by the abstraction of the state. For one of the most influential nineteenth century works evidencing the concept that states as entities are the supreme lawmaking authority and correspondingly are not subordinate to the international system, *see* JOHN AUSTIN, THE PROVINCE OF JURISPRUDENCE DETERMINED (Curwen Press 1954) (1832).

[80] *See* W. Michael Reisman, *Comment, Sovereignty and Human Rights in Contemporary International Law*, 84 AM. J. INT'L L. 866 (1990) (tracing the development from what he calls "the sovereign's sovereignty" to " the people's sovereignty").

It should be noted that this trend toward contemporary sovereignty is not unambiguous. As we have discussed, the quality of democratic governance in state-society settings leaves a great deal to be desired, and it is at least arguable that the demands of globalization are reversing the extent to which citizen voices are truly heard in the corridors of power.

The term "contemporary sovereignty" is similar but not identical to the term "responsible sovereignty" as previously used by Richard Falk. *See* FALK, *supra* note 2, at xviii. "Responsible sovereignty" denotes a notion of sovereignty in which state power is tempered and used in such a way that it promotes humane governance. We believe that contemporary sovereigns are more likely to be responsible sovereigns, but that is, of course, a subject beyond our present

hold that the ultimate source of law is the citizenry. To the extent, therefore, that one accepts states as having the sovereign power to decide whether they wish to participate in international organizations, it is because one understands them to derive this authority from the citizens in whose name they claim to act. Contemporary sovereignty's marriage of democracy and sovereignty has been increasingly accepted in the post-Cold War period. Manifesting their professed commitment to contemporary sovereignty, some states have even made democratic governance a prerequisite for their recognition of new sovereign states and to their admitting applicant states to such international institutions as the European Union.[81] While such emerging state practice indicates the increasing extent to which sovereign states need to be validated as democratic in order to be considered fully legitimate, classic sovereignty is far from extinct. There are still many nondemocratic state actors whose sovereign authority, although under some challenge, continues to be accepted.[82] As evidence of the growing belief in contemporary sovereignty, however, these states often justify their actions in the name of their people.

The world of contemporary sovereignty is a world where democratic action and sovereign action coexist as forces of legitimization. Because the citizenry rather than the sovereign is the fundamental source of political authority, citizens can bypass their sovereign intermediaries and act directly to create an international organization. Thus, either states, as representatives of

focus. For a discussion of a related connection that has engendered widespread support in recent years, *see* Michael W. Doyle, *Kant, Liberal Legacies, and Foreign Affairs*, 12 PHIL. & PUB. AFF. 205, 206-09 (1983) (marshalling empirical evidence to support the proposition that democratic states do not go to war with each other).

[81] The Bush administration made U.S. recognition of new states explicitly dependent upon democratic governance. *See* Testimony of Ralph Johnson, Deputy Assistant Secretary of State for European and Canadian Affairs, Oct. 17, 1991 , *reprinted in* FOREIGN POL'Y BULL., Nov.-Dec. 1991 , at 39, 42. Because of the lack of new states coming into existence since 1992, the Clinton administration has not significantly revisited this policy. For the recognition policies of members of the European Union, *see* European Political Cooperation, *Declaration on the "Guidelines on the Recognition of New States in Eastern Europe and in the Soviet Union,"* Dec. 16, 199 1, *reprinted in* 31 I.L.M. 1486 (1992). For a discussion of admission into the European Union, see generally Thomas Pederson, *The Common Foreign and Security Policy and the Challenge of Enlargement, in* THE EUROPEAN COMMUNITY IN WORLD POLITICS 31 (Ole Nørgaard et al. eds., 1993).

[82] At this time it is not necessary for us to discuss the question of whether elections should be considered a sin qua non for proper governance at the domestic level. For present purposes we need only establish that there is a general acceptance of electorally constituted assemblies in the world today that would serve to legitimize a GPA created directly by civil society. For the most prominent recent attempt to reconcile certain types of non-electoral systems with a conception of a "well ordered society," see generally John Rawls, *The Law of Peoples, in* ON HUMAN RIGHTS 41 (Stephen Shute & Susan Hurley eds., 1993).

their respective citizens, or more primarily, the global citizenry, acting through representative process, can create an international organization that could exercise lawmaking powers. Not only does this mean that citizens have the power to instigate a GPA, but also that if political authority does in fact ultimately reside in citizens, then logically it is the citizens themselves that have the right and perhaps responsibility to found their own assembly.

Of course, state support for the GPA should also be encouraged. As we have already discussed, their economic and other resources would be useful. Moreover, direct democratic initiative and sovereign state action are not mutually exclusive methods of legitimization. Any endorsement, financial assistance, or support from like-minded states would undoubtedly be helpful and would reinforce existing transnational grassroots efforts to achieve legitimacy. In fact, as we will next explore, the GPA would likely become empowered incrementally after it became operational. As part of this empowerment process, we will show that it could enhance its stature by persuading governments to accede to a treaty formally recognizing its legislative powers and agreeing to respect its decisions.

IV. Empowering the Assembly

If something as novel as the Global Peoples Assembly was to be introduced into this state-centric world, its democratic legitimacy would not immediately translate into formal lawmaking powers. By force of inertia, traditional power structures would remain largely intergovernmental, and the orthodox notion that international law is created only by states would pose serious conceptual and political challenges to the GPA's lawmaking powers. Given this orthodoxy, the initial legal status of the GPA would appear to be comparable to that of such nongovernmental organizations as Amnesty International, the Red Cross, and the International Olympic Committee, which have come to play significant adjunct roles in international life. In some respects, the GPA would also appear to be similar to the U.N. General Assembly, whose formal powers are mostly recommendatory. As with General Assembly resolutions, GPA resolutions and declarations initially would be treated as nonbinding in many quarters and even as usurpatory of legitimate authority structures by statist critics. As we will now explain, however, this conclusion would over time come to be seriously challenged.

A. The Socio-Political Dynamic of Empowerment

From the moment the GPA came into existence a socio-political dynamic of empowerment would be set in motion. If elections were successful, the fact that millions had participated in choosing their representatives would ensure that from its inception the GPA would have a high profile, a certain media theatricality. The political support of such an assembly would, therefore, be noticed, and, as the world's only directly elected body, it could become very useful to citizen groups as well as some governments, and other interests, wishing to legitimize their own policy objectives. Generally speaking, those who believed the GPA to be sympathetic to their position would likely seek its support. This would put those with opposing policy objectives in the position of either conceding the support of the assembly or competitively engaging its process. Of course, much would depend upon extraneous political factors as well as the quality of representation within the GPA, but it is likely that with continuing usage the GPA's importance as a center of legislative activity would grow. Interested parties would become accustomed to viewing the GPA as a place to resolve differences, and mechanisms for doing so would become established, familiar, and accepted. As the GPA became a center of activity, press coverage of its proceedings and pronouncements would expand, thereby deepening public awareness and reinforcing its influence.

Our fifty-year experience with the United Nations provides empirical evidence suggesting the potential for the GPA to become empowered in the way described above. A review of modern statecraft reveals that, as we foresee would happen in some form with the GPA, the practice has been for states to have recourse to the U.N. General Assembly or Security Council when it has served their strategic interests.[83] And, as we also anticipate would be the case with the GPA, the influence and importance of the United Nations has grown when states have chosen to employ these main political organs. Of course, the growth in its influence has not been linear. Rather it has ebbed and flowed, mostly reflecting the political will of the major governments and,

---

[83] While almost any major international political matter of the last half-century could be used to demonstrate this practice at work, one of the most striking examples is U.S. Secretary of State Dean Acheson's novel recourse to the U.N. General Assembly during the Korean War. The U.N. Charter had quite clearly assigned matters of collective security to the Security Council, allowing the General Assembly only a limited subordinate role. Unable, because of the Soviet veto, to get the Security Council to give what he considered an essential mandate to prosecute the U.S.-led war effort into North Korean territory, Acheson circumvented the Security Council and got the authorization he sought from the General Assembly. Wanting to legitimize the General Assembly's ability to play such a role when needed in the future, he eventually secured General Assembly passage of the Uniting for Peace Resolution. Through this resolution, the General Assembly essentially proclaimed for itself a much more significant role in matters of collective security than it previously had held. For a very interesting description of the events surrounding the adoption of the Uniting for Peace Resolution, *see* THOMAS FRANCK, NATION AGAINST NATION: WHAT HAPPENED TO THE U.N. DREAM AND WHAT THE U.S. CAN DO ABOUT IT 39-41 (1985).

above all, that of the United States.[84] When these governments have sought U.N. support, as President Bush prominently did in seeking Security Council authorization of the Gulf War, the tendency has been to strengthen the organization so as to undermine the legitimacy of future unauthorized actions.[85] Much of the controversy surrounding NATO's recent intervention against Serbia evidences a genuine post-Gulf War expectation that great power interventions should be authorized by the United Nations. We anticipate that the GPA's path to empowerment would be similarly influenced by such state behavior.

The GPA, however, has the potential to become more influential than the United Nations (operating without a popularly elected organ) has been thus far. First, since the GPA would be constituted without direct dependence on states, it would be less vulnerable than the United Nations to damage from the strategic oscillations of states. More fundamentally, the logic that would propel global policymakers to utilize the GPA is, in some respects, more powerful than the geopolitical calculations that intermittently motivate them to pursue the United Nations. This is because global policymakers would find the GPA, once constituted, an obvious way to compensate for the restrictions on democratic citizen participation imposed by existing international organizations, the so-called "democratic deficit."[86] At present, the sole primary constituencies to which most international organizations must respond are states, not citizens. These organizations typically afford few opportunities for citizens to participate directly through lobbying and pressure tactics. While the positions that states take in international regulatory bodies are, in varying degrees, influenced by their citizens, this influence is too attenuated and capital-driven to satisfy the conditions of a truly representative democracy. Given the strength of the northern dominated international capital forces, the people of the South are particularly unrepresented, even indirectly, in the formation of global regulatory policy.

Not only is the present international decision-making process tainted by a disregard for democratic principles, but the lack of direct democratic ac-

---

[84] For a highly personalized eyewitness account of the wavering U.S. commitment to the United Nations during his tenure as Secretary General, see BOUTROS BOUTROS GHALI, UNVANQUISHED: A U.S.-U.N. SAGA (1999).

[85] For a discussion of the effect of state commitment to the United Nations on the strengthening of the organization, see Anthony Parsons, *The UN and the National Interests of States, in* UNITED NATIONS, DIVIDED WORLD 104 (Adam Roberts & Benedict Kingsbury eds., 2d ed. 1993); and JAVIER PÉREZ DE CUELLAR, ANARCHY OR ORDER (1991).

[86] *See generally* James Crawford & Susan Marks, *The Global Democracy Deficit: An Essay in International Law and its Limits, in* RE-IMAGINIG POLITICAL COMMUNITY 72, 72-90 (Daniele Archibugi et al. eds., 1998) (discussing the normative requirements for democratic decision-making in international organizations).

countability to citizens has also significantly affected policy outcomes. The international regulatory framework has been driven almost exclusively by the neo-liberal (free-market) economic precepts so in vogue with the forces of international capital. Community interests, especially the interests of poor people, are largely ignored. The socially sensitive regulatory framework adopted by some of the more progressive societies of the twentieth century has been cast aside in favor of nineteenth century laissez-faire models.

The problem of the democratic deficit has taken on greater urgency in a world of burgeoning transnational regulation. With the rise of globalization, states increasingly find themselves forced to rely on transnational regulation to deal with matters that were previously domestic in nature.[87] Subject matter directly affecting peoples lives — ranging, for example, from the extent to which banks can extend loans to residents of poor neighborhoods,[88] to the

---

[87] For many different reasons, globalization encourages transnational regulation. Without such regulation companies are forced to manufacture to varying national product specifications, and thus cannot fully take advantage of global economies of scale. *See* Joel P. Trachtman, *International Regulatory Competition, Externalization, and Jurisdiction*, 34 HARV. INT'L L. J. 47, 66-67 (1993). In addition, states competing for investment capital are continuously forced to relax "anti-business" regulations in a "race to the [regulatory] bottom." For a theoretical understanding of this dynamic, see DREW FUDENBERG & JEAN TIROLE, GAME THEORY 9-10 (1992). Environmental measures, labor standards, and the prudential regulation of capital markets are all imperiled. For a relevant discussion of environmental regulation, see generally Daniel C. Esty, *Revitalizing Environmental Federalism*, 95 MICH. L. REV. 570, 638 (1996). For a relevant discussion of labor regulation, see generally Virginia A. Leary, *Workers' Rights and International Trade: The Social Clause* (GATT, ILO, NAFTA, U.S. Laws), 2 FAIR TRADE & HARMONIZATION 177, 183 (Jagdish N. Bhagwati & Robert E. Hudec eds., 1996). For a relevant discussion of banking regulation, see generally RICHARD DALE, REGULATION OF INTERNATIONAL BANKING 172-85 (1986). Even the effective enforcement of criminal law and the coordination of civil litigation demand transnational regulation. See Andrew Strauss, *A Global Paradigm Shattered: The Jurisdictional Nihilism of the Supreme Court's Abduction Decision*, 67 TEMP. L. REV. 1209 (1994) (discussing the need for a globally coherent allocation of the authority to enforce criminal laws); Andrew Strauss, *Beyond National Law: The Neglected Role of the International Law of Personal Jurisdiction in Domestic Courts*, 36 HARV. INT'L L.J. 373 (1995) (discussing the need to internationally coordinate rules of civil jurisdiction).

[88] Minimally capitalized banks have a competitive advantage in loaning money but also have a heightened risk of failure. Because of the interconnectedness of the international banking system, failures by such banks have the potential to imperil the whole of the world economy. In an attempt to ameliorate this problem, the U.S. Federal Reserve and other central banks signed an accord establishing minimum capital adequacy standards. *See* Basle Committee on Banking Regulations and Supervisory Practices, *International Convergence of Capital Measurement and Capital Standards, reprinted in* 51 BNA'S BANKING REP. 143 (July 25, 1988). Under this Accord, bank loans are distinguished based upon their relative risk. The riskier a bank's loan portfolio, the greater the capital the bank is required to hold to protect against potential defaults. The Accord adversely affects the availability of loans for lower income housing in the United States because, under the Federal Reserve's interpretation, loans

length of time patents run,[89] to food safety[90] — are now, at least in part, transnationally regulated. Such regulation is likely to continue to grow for the simple reason that, even at present, the degree of regulation is not nearly sufficient to address the instabilities associated with the activities of global capitalism.

Governing elites within both national governments and international organizations have begun to worry that the democratic deficit is undermining the legitimacy of the present international system. Their concern is that growing popular resistance to a system that denies citizen participation will make it increasingly difficult to implement effective transnational regulation. These concerns were heightened recently when noisy street protests devoted in part to challenging the anti-democratic nature of international economic

for multi-family dwellings, common in low-income neighborhoods, are considered quite risky. *See generally* Duncan E. Alford, *Basle Committee International Capital Adequacy Standards: Analysis and Implications for the Banking Industry*, 10 DICK. J. INT'L L. 189 (1992). *See also Deposit Insurance Reform and Financial Modernization, 1990: Hearings on Reforming Federal Deposit Insurance, Modernizing the Regulation of Financial Services, and Maintaining the International Competitiveness of U.S. Financial Institutions Before the Senate Comm. on Banking, Housing, and Urban Affairs*, 101st Cong. 157 (1990) (testimony of Jane Uebelhoer, Legislative Director, ACORN).

[89] The Trade Related Intellectual Property Rights Agreement that came out of the Uruguay Round of international trade negotiations requires that the term of patent protection granted by state parties be at least twenty years. See Agreement on Trade-Related Aspects of Intellectual Property Rights, Part 11, 35, art. 33. This was three years beyond the prior level of patent protection in the United States.

[90] Under the Agreement on the Application of Sanitary and Phytosanitary Measures, World Trade Organization members have the right to take sanitary and phytosanitary measures that are "necessary" for the protection of human and animal health. See Agreement on the Application of Sanitary and Phytosanitary Measures, *reprinted in* GATT Secretariat, *The Results of the Uruguay Round of Multilateral Trade Negotiations: The Legal Texts* 69, at 70 art. 1 (1994). Crucially, however, such measures must, under the language of the agreement, be based on "scientific principles" and "sufficient scientific evidence." *Id.* at art. 2.2. Recently, the WTO Appellate Body found that a European Union regulation banning (for health reasons) the domestic selling of hormone-fed beef was not based on scientific evidence. See EC Measures Concerning Meat and Meat Products (Hormones), WT/S26/R/USA (Aug. 18, 1997). For further discussion of the Sanitary and Phytosanitary Agreement, see John J. Barcelo, *Product Standards to Protect the Local Environment - the Gatt and the Uruguay Round Sanitary and Phytosanitary Agreement*, 27 CORNELL INT'L L.J. 755 (1994). International food safety standards are also heavily influenced by Codex, an intergovernmental organization whose mission is to harmonize food safety standards. For a discussion of Codex's role in the setting of international food safety standards, see generally Lewis Rosman, *Public Participation in International Pesticide Regulation: When the Codex Commission Decides, Who Will Listen?*, 12 VA. ENVTL. L. J. 329 (1993). The trend is in the direction of increased international regulation in the food safety area. See Jim Hoagland, *Europe's Food Fright*, WASH. POST, June 24, 1999, at A27 (reporting on proposal by French President Jacques Chirac for a world scientific authority that would certify food safety).

decision-making broke out at the December, 1999 World Trade Organization ministerial conference in Seattle, and at the April, 2000 annual meeting of the International Monetary Fund and World Bank in Washington D.C.[91] Even before Seattle, this problem was beginning, for example, to be taken more seriously within the World Trade Organization (WTO), where those concerned with promoting the organization[92] have recognized that some type of democratic process is needed to counter growing popular opposition to many of its initiatives.[93]

---

[91] For a very good discussion of the social concerns and political alliances behind the Seattle protests, see David Moberg, *For Unions, Green's Not Easy*, THE NATION, Feb. 21, 2000, at 17. For a discussion of the new activism generally, see Barry Came et al., *People Power*, MACLEAN'S, Jan. 1, 2000, at 220.

[92] U.S. President William Clinton has declared that "we must modernize the WTO by opening its doors to the scrutiny and participation of the public." President William Clinton, Remarks at the World Trade Organization in Geneva, Switzerland, 1 PUB. PAPERS PRESIDENT 807, 810 (May 18, 1998). *See also* E.J. Diome, Jr., *Clinton Seeks Leveling Up Instead of Down*, INT'L HERALD TRIB., June 4, 1998, at 10 (discussing the policy pressures that brought Clinton to make the address).

Similarly, former WTO Director General Renato Ruggiero has stated that "[c]onsensus does not just mean agreement among governments. Consensus also means dialogue with our citizens" and that he "intend[s] to devote a great deal of [his] time to improving this dialogue-a dialogue including the widest possible representation and transparency in all the activities of the WTO." Director General Renato Ruggiero, Address at the Friedrich-Ebert-Foundation Hamburg, Germany (June 11, 1998). in ABI/INFORM. While Ruggiero's comments on the subject of democratizing the WTO were more cautious than President Clinton's, they make clear that he felt it necessary to address the organization's democratic deficit. Ruggiero followed up shortly after making these comments by beginning a program of regular briefings for NGOs on the work of WTO committees and working groups, and by disseminating documents, position papers, and newsletters submitted by NGOs to the organization's member states. See World Trade Organization, Press Release 10, *Ruggiero Announces Enhanced WTO Plan for Cooperation with NGOs*, July 17, 1998.

[93] Popular opposition to the organization can be seen in many comers. In the United States, for example, presidential candidates as different as Ralph Nader, Pat Buchanan, and Ross Perot have made opposition to the WTO central to their message. Nader has heavily criticized what he calls "the GATT and NAFTA systems of autocratic governance." RALPH NADER & WESLEY J. SMITH, NO CONTEST: CORPORATE LAWYERS AND THE PERVERSION OF JUSTICE IN AMERICA 338 (1996). For further elaboration on Nader's views on the WTO and international trade generally, see Andrew Strauss, *From GATTzilla to the Green Giant: Winning the Environmental Battle for the Soul of the World Trade Organization*, 19 U. PA. J. INT'L ECON. L. 769, 771-72 (1998). Patrick Buchanan, criticizing from the right-wing of the ideological spectrum as he does, argues that the United States should leave the organization so as to preserve the country's national sovereignty, but the appeal of his argument is clearly aided by the fact that the WTO is run without democratic accountability by what he refers to as "nameless, faceless foreign bureaucrats." PATRICK J. BUCHANAN, THE GREAT BETRAYAL: HOW AMERICAN SOVEREIGNTY AND SOCIAL JUSTICE ARE BEING SACRIFICED TO THE GODS OF THE GLOBAL ECONOMY 313 (1998). *See also* ROSS PEROT WITH PAT CHOATE, SAVE YOUR JOB, SAVE OUR COUNTRY: WHY NAFTA MUST BE STOPPED-NOW (1993) (focusing on the then-

But what to do? The ad-hoc response has been gradually to open international regulatory bodies to participation by NGOs and citizens' associations. Even if implemented in a way that includes voices from the South, which is unlikely, such an approach cannot solve the problem of the democratic deficit. As industrial and other narrow private interests, as well as eccentric fringe groups, increasingly acquire NGO identities, international organizations like the WTO will show themselves institutionally incapable of impartially overseeing the process of representation. They are unlikely to be perceived as fair in deciding which organizations should be allowed to represent the global citizenry in decision-making and which should not.[94] To date, this problem has been avoided only because NGOs and citizens' associations have had either very little, or informal and episodic, influence. There is currently no viable strategy for overcoming the democratic deficit.

Imagine, therefore, what could happen if the GPA were to appear upon the global stage. Governing elites would be offered an attractive vehicle to help overcome popular resistance — an offer that would likely be very difficult to disregard. After all, civil society organizations promoting global regulations of an equitable character would themselves most likely see the GPA as a promising tool to help overcome resistance by states and private sector interests to regulation on behalf of the public good. In a specific application of the theme we have already suggested, if governing elites passed up the offer to enter the GPA's arena to find legislative common ground and allowed their challengers to go unanswered, their democratic deficit problem would be compounded. Not only would their approach to global public policy derive from a non-democratic process, but in addition their attempt to regulate would actually appear to defy the one body capable of genuinely speaking on behalf of the people of the world.

When the full dynamic is understood, not only would the GPA be a forum specifically created to give citizens legislative standing, but it is also hard to imagine that those involved with the established international regulatory

---

impending congressional debate on NAFTA but criticizing the GATT (soon to become the WTO) as well). Similar opposition to the organization and its initiatives is present in many other countries.

[94] David Rieff has articulated some of the difficulty inherent in discriminating between civil society organizations:

> Why, for example, is the International Campaign to Ban Landmines viewed as an exemplar of civil society instead of, say, the National Rifle Association, which whatever one thinks of its politics, has at least as good a claim to being an authentic grassroots movement? The UN bitterly resisted having to recognize the NRA as a legitimate NGO. And yet if we think of NGO as a description and not a political position, the NRA obviously qualifies.

Rieff, *supra* note 64, at 15.

institutions could afford to disregard the GPA.[95] In fact, in a likely portent of a future with the GPA, European Union policymakers, as discussed, have begun to seek legislative legitimacy by attempting to secure the imprimatur of a democratically elected assembly. Having become acutely aware that the democratic deficit is undermining the legitimacy of directives promulgated by the unelected and "faceless" European Commission, successive European Union treaties have strengthened the powers of the popularly elected European Parliament.[96] Like their European counterparts, global policymakers would likely find a popularly elected assembly a very helpful way to shrink the democratic deficit, or at least such hopeful evolution should not be dismissed in advance.

---

[95] Perhaps Professor Franck best summarized the political calculus that is likely to bring the global regulators into the realm of the GPA when he pragmatically concluded that "consent benefits the governing as much as the governed," in that it helps to secure the "habitual voluntary compliance of its subjects." Thomas Franck, *The Emerging Right to Democratic Governance*, 86 AM. J. INT'L L. 46, 48 (1992).

[96] *See supra* notes 65-71 and accompanying text; see also Philippe Manin, *The Treaty of Amsterdam*, 4 COLUM. J . EUR. L. 1, 11-14 (1999) (discussing how certain of the European Union treaties and particularly the Treaty of Amsterdam have strengthened the European Parliament).

International elites have been very concerned about the implications of the democratic deficit for governance within the European Union and about how to remedy it. For example, George Soros, one of the most high-profile advocates for global financial regulation, has observed that "[w]hat the people see is a top-heavy bureaucratic organization that works in convoluted ways shrouded in secrecy and not responsible to the public," and that "[t]o change this perception, the administration ought to be made more directly responsible to the people, either through the national parliaments or the European Parliament." GEORGE SOROS, THE CRISIS OF GLOBAL CAPITALISM: OPEN SOCIETY ENDANGERED 228 (1998).

Likewise, in explaining the increased power given to the European Parliament in the Treaty of Amsterdam, Michel Petite, Director of the Secretariat General of the European Commission has commented:

> The call for. . . a more democratic functioning of the [European Union] institutions, could be heard from many sources. This demand came not only from the European Parliament, but also from almost all Member States. From the Constitutional Court in Germany, various nongovernmental organizations, and simply good natural sense. Naturally, the Commission also sought a more democratic system of institutional structures and operations. It was absolutely clear that any extension of the Community order would require a more proper and classically democratic system. Failing this, we would undoubtedly face major constitutional problems in Member States and eventually bring the European Construction to a halt.

Michel Petite, Essay, *The Commission's Role in the ZGC's Drafting of the Treaty of Amsterdam*, 22 FORDHAM INT'L. L. J. 72, 76 (1999). It should also be noted that during European Union treaty negotiations the Parliament has been very skilled at winning expanded powers by using its existing powers for negotiating leverage.

B. The Ideological Dynamic of Empowerment

To the extent that this socio-political dynamic occurs, ideological doctrine proclaiming the GPA's authority to create binding law would correspondingly gain gradual acceptance. What we call an ideological dynamic of empowerment would likely be set in motion as observers sympathetic to global democracy, as well as those with political or economic interests in promoting the GPA, began to fashion formal legal arguments as to why its resolutions should be considered binding.

In a world that largely subscribes to principles of contemporary sovereignty, the argument that the GPA — the only body to represent the peoples of the world — has the power to create binding transnational law presumably would have a wide acceptance at grassroots levels. Law professors would write law review articles. Private litigants would (when helpful to their cause) invoke the authority of the GPA before domestic tribunals. Even some state parties arguing in international fora would do likewise when they found it useful in making their own legal cases. Over time, it is probable that independent-minded judges would be inclined to accept such arguments.[97] Such acceptance would, of course, have a powerful spiraling effect leading more such arguments to be made. Some progressive governments might even come to assert that the GPA's resolutions should be considered binding law both internationally and within their countries. These multifaceted developments would help erode adherence to the orthodox legal doctrine that binding international norms can be created only by states.[98]

---

[97] In fact, in the classic case of *McCulloch v. Maryland*, the U.S. Supreme Court used the argument that a popularly elected assembly has the inherent power to impose binding law on "sovereign" states to justify its expansive understanding of federal power. See McCulloch v. Maryland, 17 U.S. (4 Wheat.) 316 (1819). In *McCulloch*, U.S. states claimed that the U.S. Constitution emanated from their independent sovereignties, and that the exercise of federal power could not predominate over the states' claims to power. Writing for the Court, Chief Justice Marshall rejected this theory, asserting that the federal government's democratically constituted legislative powers came directly from the people, not the states and, therefore, that the states could not themselves limit the grant of power to Congress. McCulloch, 17 U.S. (4 Wheat.) at 403-07.

Even in the absence of the GPA, the World Court is cautiously beginning to recognize the legal relevance of the international citizenry. In particular, Judge Weeramantry in his dissent to the *Nuclear Weapons Advisory Opinion* provides a broad jurisprudential argument in support of the role of the people in reshaping international law on matters of security and survival. *See* Richard Falk, *The Nuclear Weapons Advisory Opinion and the New Jurisprudence of Global Civil Society*, 7 TRANSNAT'L. L. & CONTEMP. PROBS. 333, 345-47 (1997) (discussing Judge Weeramantry's dissent).

[98] Of course, statist- and capital-driven resistance to such developments could also be anticipated.

Again, our experience with the United Nations can give us some confidence in the ultimate ideological empowerment of the GPA. Under the U.N. Charter, the powers of the General Assembly (which gives each state one vote) are largely precatory.[99] Despite this fact, in the 1960s and 70s, developing countries (that constituted and still constitute a majority of the Assembly) and their academic supporters developed a variety of legal theories to advance the argument that the resolutions of that organ should be considered binding.[100] Powerful geopolitical forces led by the United States kept these theories from ever gaining primacy, although they did gain a measure of acceptance.[101] Though it is difficult to predict, arguments supporting the

[99] *See* U.N. CHARTER arts. 10-15. *But see* U.N. CHARTER arts. 17, 85.

[100] This is an area that has engendered a great amount of scholarly discussion. For some of the major works, see Bin Cheng, *Custom: The Future of General State Practice in a Divided World, in* THE STRUCTURE AND PROCESS OF INTERNATIONAL LAW: ESSAYS IN LEGAL PHILOSOPHY DOCTRINE AND THEORY 513, 531 (R. St. J. Macdonald & Douglas M. Johnston eds., 1986) (arguing that general international law is formed when " [t]he state[s] concerned accept[ ] that the norm in question is of a legal character. . . and, therefore as such, carries legal rights and duties erga omnes"); THEODORE MERON, HUMAN RIGHTS AND HUMANITARIAN NORMS AS CUSTOMARY LAW 87-88 (1989) (suggesting that, "[t]he passage of norms agreed upon in international conferences into customary law through the practice, including the acquiescence, of states constitutes a common, generally accepted method of building customary international law."); Oscar Schachter, *International Law in Theory and Practice*, 178 RECUEIL DES COURS 110, 111 (1982) (arguing that the " formative influence" held by General Assembly resolutions "in developing international law is a "natural consequence" of the General Assembly's being "the central global forum for the international community"); Christopher C. Joyner, *U.N. General Assembly Resolutions and International Law: Rethinking the Contemporary Dynamics of Norm-Creation*, 11 CAL. W. INT'LL.J. 445 (1981) (discussing the influence of the General Assembly on general international law); Rosalyn Higgins, *The Role of Resolutions of International Organizations in the Process of Creating Norms in the International System, in* INTERNATIONAL LAW AND THE INTERNATIONAL SYSTEM 21, 28 (William E. Butler ed., 1987) (arguing that the norm creating ability of resolutions depends on a number of factors, such as "the majorities supporting their adoption"); Remarks of Judge Jimenez de Arechaga, *in* CHANGE AND STABILITY IN INTERNATIONAL LAW-MAKING 48, 48 (Antonio Cassese & Joseph H. Weiler eds., 1988) (arguing that the General Assembly, where all states are represented, is a forum wherein rules of international law are generated by consensus). CJ: Jonathan Charney, *Universal International Law*, 87 AM. J. INT'L L. 529, 544 (1993) (noting that developments in international law "often get their start or substantial support from . . . resolutions . . . debated in [ ] forums [like the General Assembly]"); Richard Falk, *On the Quasi Legislative Competence of the General Assembly*, 60 AM. J. INTL. L. 782, 785 (1966) (observing "a discernible trend from consent to consensus as the basis of international legal obligations").

[101] *See* e.g., *Filartiga v. Pena-Irala*, 630 F.2d 876, 883 (2d Cir. 1980) (citing with approval the observation that the General Assembly's Universal Declaration of Human Rights "no longer fits into the dichotomy of 'binding treaty' against 'non-binding pronouncement', but is rather an authoritative statement of the international community"). *See also* Louis Henkin, *Resolutions of International Organizations in American Courts, in* ESSAYS ON THE DEVELOPMENT OF THE INTERNATIONAL LEGAL ORDER 199, 205 (Frits Kalshoven et al. eds., 1980) (arguing that General Assembly resolutions "might be given effect by [U.S] courts without

authority of a popularly elected assembly to create binding international law should be significantly more persuasive to democratic ears than have been arguments supporting the legal powers of the General Assembly. In fact, such arguments could become particularly compelling if the GPA chose to put the question of whether it should have binding powers to a vote of the global citizenry in an international referendum. If approved pursuant to a credible procedure, the fact that the GPA's authority to create binding law was granted explicitly by those representing the global citizenry would be a powerful argument in favor of its prerogatives.

C. State Acceptance of the Global Peoples Assembly

These empowering socio-political and ideological dynamics would in themselves be unlikely to settle the issue within the international community of whether the GPA had the authority to create binding law. Rather, their most significant role would be to help create a political climate that might engender a definitive empowering event. Such an event could occur if states were to validate the GPA in a formal way, such as by treaty accepting and specifying the importance of its role within the overall constitutional structure of the U.N. system of global governance. In the wake of such a validating process, the GPA's legal authority would be solidified, and the controversial character of its lawmaking claims would be overcome.[102]

Is it reasonable to expect states to give their blessings to the GPA? We realize, of course, that most states are not yet ready to accept any such law-

---

any intervening legislative or executive implementation" as elaborations of what could be considered to be self-executing U.N. Charter provisions). National courts in countries other than the United States have also relied on U.N. General Assembly resolutions for their legal significance. *See generally* CHRISTOPH C. SCHREUER, DECISIONS OF INTERNATIONAL INSTITUTIONS BEFORE DOMESTIC COURTS (1981).

International tribunals have been more inclined than national courts to ascribe some measure of lawmaking authority to General Assembly resolutions. In the Nicaragua case, for example, the International Court of Justice referred to General Assembly resolutions as evidence of the international law on use of force and nonintervention. See Military and Paramilitary Activities (Nicar. v. U.S.), 1986 I.C.J. 14, 98-107 (June 27). In the South West Africa and the Western Sahara cases, the International Court of Justice gave legal effect to General Assembly declarations on self-determination and independence of peoples in territories that have not yet attained independence. *See* Legal Consequences for States of the Continued Presence of South Africa in Namibia (South-West Africa) Notwithstanding Security Council Resolution 276 (1970), 1971 I.C.J. 16 (June 21); see also Western Sahara, 1975 I.C.J. 12 (Oct. 16). For a reference to the influence of General Assembly resolutions relating to sovereign immunity and economic development on the decisions of international arbitrators, see Georges R. Delaume, *Economic Development and Sovereign Immunity*, 79 AM. J. INT'LL. 319 (1985).

[102] Of course, new dangers of co-optation by certain states or private interests attempting to distort the democratic process might have to be confronted.

making entity and that civil society must necessarily take the lead in creating the GPA. Over time, however, the twin dynamics of socio-political and ideological empowerment would likely enlarge the realm of the possible, especially as the dynamics of globalization create more and more of a one-world awareness among peoples everywhere.

As the GPA grew in informal influence and increasingly came to act as a de facto legislature, the global structure of power would gradually reconfigure, and corresponding attitudes among state actors (with some possible exceptions) would tend to become more accepting. States would only be called upon to recognize legally a transformation that was already occurring. Along with this, the assembly's high profile and democratic legitimacy would give it a powerful ability to lobby governments effectively on its own behalf. Finally, in determining whether states would eventually come formally to concede power to such an assembly, it is important not to exaggerate the depth of present resistance of many states to such a move. After all, as we have discussed, the European Union countries through the European Parliament have been experimenting with a popularly elected transnational assembly for some time, and have consistently acted to strengthen its role, despite the consequent weakening of traditional sovereign prerogatives.

V. Conclusion: Assessing the Organizational Challenge

We do not mean in any way to minimize the logistical difficulties that would be involved in creating a Global Peoples Assembly that would have a major impact on world order. The establishment of electoral districts throughout the world would be necessary; global voter rolls would have to be generated; a system of campaign finance and other election rules would need to be established; and attempts to manipulate or undermine elections would have to be effectively guarded against.[103] Initially, some governments would not allow elections to occur in their countries, or at least not on acceptable terms. Until sufficient pressure could be brought to bear by transnational democratic

---

[103] Actually, global civil society has a good deal of experience in the area of election monitoring. Organizations such as the Inter-Parliamentary Union, the Carter Center, and the Swedish Organization, International Democratic Elections Assistance - to name but a few of the more prominent ones – have been heavily involved in this endeavor for a number of years. *See generally* GUY S. GOODWIN-GILL, FREE AND FAIR ELECTIONS: INTERNATIONAL LAW AND PRACTICE (1994) (providing on behalf of the Inter-Parliamentary Union a practical guide for assessing what constitutes free and fair elections); W. Michael Reisman, *International Election Observation,* 4 PACE U. L. SCH. Y.B. INT'L L. 1, 6-7 (1992) (discussing the role that N W observers have played in international election monitoring).

forces, citizens of these countries would have to go unrepresented, or possibly be represented by delegates selected in some other way. Once the GPA was constituted, meeting facilities, translation services, and staff and other support services would have to be arranged. Undertaking arrangements on this scale would be organizationally daunting, not to mention very expensive.[104]

Sufficient commitment, however, can overcome logistical difficulties. If enough individuals, advocacy groups, foundations, churches, labor unions, and other organizations with resources were to get behind the GPA, the task could be accomplished — especially with the help of supportive governments.[105] What is remarkable, given the tremendous potential for such an assembly truly to transform global governance, is that in certain ways it is politically less challenging than what global civil society already accomplished in the 1990s. Unlike those previous projects, which relied upon interstate treaty arrangements, there is little (at least within the more democratic societies) that opponents can do to thwart the GPA's development if popular support is forthcoming. The tremendous clout that many opponents of progressive international reform enjoy within national governments, and that can be used to block adherence to treaty regimes, will be of little use to them. They will have little option other than to criticize the project and refrain from participating in it.

How tremendously energizing it will be when people of democratic spirit and ethos in the world realize the full promise of this venture. After the great suffering of the last century, it would provide an auspicious beginning of this new millennium to have a vision as bold as the GPA put forth in a serious manner that captured the imagination of many people. Even before the ultimate goal of a GPA could be achieved, its very emergence on the international agenda would give a concrete and positive vision around which those who have voiced their objections to the anti-democratic nature of global institutions such as the WTO, the IMF, and the World Bank could organize. Independent of the specific merits of a global popularly elected assembly as the ultimate solution to the democratic deficit, the general cause of furthering the democratization of the global order could only be aided by the advance-

---

[104] There is reason to believe, however, that emerging technologies could help overcome certain logistical barriers. *See Casting Ballots Through the Internet*, N.Y. TIMES, May 3, 1999, at C4 (reporting on start-up companies that are developing systems to enable voters to cast ballots over the Internet).

[105] *See* Andrew Strauss & Richard Falk, *For a Global Peoples Assembly*, INT'L HERALD TRIB., Nov. 14, 1997, at 8 (suggesting that individuals with the resources and inclinations of George Soros and Ted Turner have the capacity to help make a Global Peoples Assembly a reality).

ment of a concrete proposal around which media and public attention could be focused.

The analysis we have presented explaining why a citizen-organized assembly would be viable has looked backward at the dynamics of empowerment from the world as it exists. What we have left out of consideration is the tremendous transformative energy that could be unleashed when those people who have found their deep-seated aspirations for a better world so difficult to realize are presented with a viable vision of a GPA around which they can unite. Once unleashed, this energy for innovation and change could be as infectious as the negative energy of despair and hate that have so often acted to constrain our future. If this energy of innovation were to spread, the political terrain that has previously commanded adherence beneath the banner of "realism" would shift in supportive ways. If this political awakening occurs, the establishment of the Global Peoples Assembly might come to be seen as but one of several giant steps down the path leading to the emergence of humane governance for all the peoples of the world, thereby also fulfilling the quest for a form of world order that incorporates the ideas and practices of global democracy.

On the First Branch of Global Governance

by Andrew Strauss

Widener Law Review, 2007[*]

I. Introduction

Because leaders can't lead if followers won't follow, the political character of societies are determined by how citizens collectively choose to condition their willingness to follow. They may reflexively follow the commands of those proclaiming authority or they may subject their obedience to various forms of moral scrutiny. In the 20th century, successfully demonstrating this reality, Mohandas Gandhi, Martin Luther King, Vaclav Havel and others overturned political orders by inspiring millions to condition their willingness to follow on specific normative criteria.

Citizens do not make collective determinations of when they will follow on an *ad hoc* basis. In order to function effectively, societies need to institutionalize the coordination of actions and no society has yet figured out how do so without delegating at least limited decision-making powers. At the level of organizing global society, citizens on matters of common global interest have almost exclusively come to accept the obligation to follow the legal commands of national authorities as opposed to global authorities. And, those in democratic societies have not demanded the same opportunity to participate in global decision-making that they have come to expect in national decision-making.

The international institutional configuration that has resulted does not measure up to contemporary standards of fairness and democracy. Certain national authorities dominate the global system, and citizens with access to those national authorities can leverage that access into tremendous influence over the course of international events. Average citizens in less powerful countries are, in contrast, largely powerless to influence global events. This stark inequality of power has lead to pervasive feelings of exclusion, anger

[*] Reprinted from WIDENER LAW REVIEW, Volume 13, No. 2, 2007.

and alienation and has no doubt become a major contemporary cause of global tensions.

The global situation would be very different if global citizens all had an equal vote in selecting global authorities whose legal commands they came to follow on matters of common global concern. Then at least at the level of formal political structure, all citizens would have a fair say in decision-making. What I am describing, is almost universally believed to be intrinsic to domestic democracy, a popularly elected parliament, or what, in Montes-quieu's terms, might be thought of as the first branch of global governance.

Richard Falk and I have long argued for a global parliament. In this article I wish to address specifically how such a global institution might be practically instituted given present-day political realities. I compare four approaches for bringing about such an organization. If any of the four could be successful, the democratic principle that citizen followership should be conditioned upon democratic leadership would be for the first time intro-duced into the global system.

II. Four Approaches, But One Common Principle

The one principle common to the four approaches to creating a Global Par-liamentary Assembly (GPA) is that the parliament begins initially as a large-ly advisory body rather than a full-fledged legislative assembly with binding powers. The most successful example of a popularly elected transnational parliament is the European Parliament, and that institution of the European Union started in the early days of European integration with only advisory powers. Today, half a century later, it has attained for itself a considerable role in European Union lawmaking[1] and there continue to be important pro-posals for further strengthening its powers.[2]

---

[1] Since 1970 the European Parliament has, in stages, gained a considerable role in legis-lative decision making within what is today the European Union. The vast majority of deci-sions now require approval by the Parliament. *See* Wilhelm Lehmann, *European Parliament Fact Sheets 1.3.1,* http://www.europarl.europa.eu/facts/1_3_1_en.htm; Wilhelm Lehmann, *European Parliament Fact Sheets 1.3.2,* http://www.europarl.europa.eu/facts/1_3_2_en.htm.

[2] The proposed European Constitution (the ratification of which was suspended follow-ing its defeat by voters in France and the Netherlands) gave considerably enhanced powers to the European Parliament. The growing number of countries in the European Union is making it increasingly difficult for national leaders to reach the degree of consensus required for decision-making. Consequently, suggestions are being forwarded that important decisions can be better made by the European Parliament. *See, e.g.,* Europe Info. Serv., *Future of Europe: Reinforced Cooperation Cannot Be Excluded as an Option*, EUROPEAN REPORT, Feb. 9, 2007 (available via subscription; on file with author).

What we can draw from the experience with the European Parliament is that postponing the day in which the GPA will have significant legislative powers enhances its political viability because it encourages those who are presently powerful to focus more on the organization's abstract neutral benefits than on how it might negatively impact their short-term political interests. Of particular importance, this would apply to political leaders who would largely be ceding to the parliament their successors powers, rather than their own.

Starting the GPA in a modest way as an advisory body, as the price to pay for initial political viability, does not mean that the parliament could not come to play a central role in global governance. Even in its early days, despite the parliament's lack of formal powers, it would be well positioned to bring about some measure of accountability to the global system. If, for example, it suspected that various international organizations were engaged in malfeasance or nonfeasance, it could hold hearings and issue reports. By virtue of being the one popularly elected body at the global level, its reports would carry the weight of moral authority and have a strong claim to be taken seriously. And over the long term it is likely that the parliament would grow far beyond its relatively humble beginnings. By virtue of its democratic legitimacy it would be poised to assume the sorts of powers that are typical of parliaments all over the world.

Citizen groups would likely seek to have the parliament's moral authority associated with their cause. For example, groups critical of existing international economic organizations such as the International Monetary Fund, the World Trade Organization and the World Bank would very likely petition the parliament to condemn various policies of those organizations. Representatives of those organizations, or other interest groups with contrary positions, are not likely to concede the legitimacy of the only popularly elected global body. Rather, as is the case with national parliaments, the global parliament would provide a political forum where the various interests would come together, and through the intermediation of their elected representatives, work out legislative compromise that could be formally and finally agreed upon by the parliament. The likely result is that these global interests would come to have a sense of ownership in the parliament, its processes and outcomes.

As the planet's organized citizenry began to reconfigure itself beyond the limitations of separate and discreet orbits around national parliaments into a new common orbit around a GPA, over time the parliaments formal powers would likely come to reflect this new political reality. Not only would the organized citizenry be inclined toward supporting the legal force of legislative results that were fashioned in response to their input, but an existing parliament could powerfully lobby governments on behalf of expanding its own powers. In a world where democratic elections have become the litmus test for legitimate governance at the local, provincial and national levels, the parliament's claim to exercise increasing authority in the name of the global citizenry would be hard to resist.

Also furthering the parliament's gradual increase in powers would be its own elected representatives. Presently there is no global institution whose constituency is the world's citizenry. The United Nations is a society of states, and the institutional loyalties of the representatives of those states are oriented toward championing the prerogatives of their states, even at the expense of an organizationally successful United Nations. The sole institutional affiliation of those elected directly to the parliament would be the parliament, and their professional status would be tied directly to the growth and empowerment of the parliament.

Perhaps the greatest reason to believe that the parliament could expand its powers within the international system is that the system is missing a large component of the machinery of effective governance. The international system has no center. There is no body to coordinate, harmonize or oversee the balkanized global bureaucracy. International organizations, whether they deal with health, labor, trade, weapons, or other matters, are all separate bodies, often created by independent treaties. Because it is well established in the popular imagination that parliaments oversee agencies of government, and there is no other candidate to play this role at the global level, the GPA would be poised to take its central place in the international system as the vertical link to the citizenry and the horizontal link between the various international organizations.

Proclaiming that an initially advisory body could incrementally, through the power of popular legitimacy, become an important international organization begs the question of how even an initially modestly empowered body could be conceived in today's world. This is the threshold question to which I will now turn. Of the four alternative approaches I will consider, the first and perhaps most obvious one is to amend the United Nations Charter to create a parliament as part of the United Nations. The second approach is for the General Assembly of the United Nations to create the parliament pursuant to its powers under the United Nations Charter to establish "subsidiary organs." The third approach is for civil society on its own initiative to create the parliament outside of official United Nations or interstate treaty processes. Finally, the fourth approach is for willing states to enter into a stand-alone treaty creating the parliament.

III. The Four Approaches to Establishing a GPA

A. Amendment of the United Nations Charter

Pursuant to Article 108 of the United Nations Charter, amendments to the Charter require approval by a two-thirds vote of the United Nations General Assembly and subsequent ratification by two-thirds of the members of the United Nations, including all of the permanent members of the United Nations Security Council.[3] Article 109 of the Charter somewhat less onerously allows for a Charter review conference to be established by a two-thirds vote of the General Assembly and an affirmative vote of any nine members of the fifteen-member Security Council.[4] Any alteration of the Charter coming out of the review conference, however, must similarly be approved by two-thirds of the conference and ratified by two-thirds of the United Nations membership including all of the permanent members of the Security Council.[5]

Amendment of the United Nations Charter pursuant to Articles 108 and 109 provide what might be called the classical route to creating a GPA. This was the approach adopted by early world federalists such as Louis Sohn and Grenville Clark in their 1958 book *World Peace Through World Law*, which includes an elected parliament as part of their scheme to turn the United Nations into a limited world government.[6] While the currents of historical change are not always predictable, the political barriers that are likely to stand in the way of such an approach would appear formidable. The mixed results of the United Nations' most recent experience in 2005 with significant reform showed just how politically difficult the U.N. reform process can be.[7] Even getting a proposal for a parliament on the United Nations reform agenda would be a difficult task. For example, neither of the two reports written for the Secretary General in advance of the 2005 reforms—the *Report of the Panel of Eminent Persons on United Nations Civil Society Relations*[8] and the *Report of the Secretary General's High Level Panel on Threats, Challenges and Change*[9]—mentioned an elected chamber of the United Nations.

Also, significantly none of the 2005 reforms, which have been implemented, have required amending the United Nations Charter. Clearly, con-

---

[3] U.N. Charter art. 108.

[4] *Id.* at art. 109, para. 1.

[5] *Id.* at art. 109, para. 2.

[6] GRENVILLE CLARK & LOUIS B. SOHN, WORLD PEACE *THROUGH* WORLD LAW: TWO ALTERNATIVE PLANS xlii-xliii (3d ed. enlarged 2d prtg. 1967) (1958).

[7] *See* U.N. GAOR, 60th Sess., U.N. Doc. A/60/355 (Sept. 14, 2005).

[8] *See* U.N. GAOR, 58th Sess., U.N. Doc. A/58/817 (June 11, 2004).

[9] U.N. GOAR, 59th Sess., U.N. Doc. A/59/565 (Dec. 2, 2004).

vincing two-thirds of the organization's membership to approve amending the Charter to create a parliament would not be easy, and ratification by that number of states would even be more difficult. Finally, securing the affirmative votes of all of the veto-welding members of the Security Council, given the reluctance of some of these countries to support progressive international initiatives, would likely be quite difficult. Perhaps, however, as Joseph Preston Baratta has suggested in *The Politics of World Federation,* the permanent member veto would not have to be the final word. He finds inspiration in the observation that the delegates to the United States Constitutional Convention of 1787 provided for ratification by nine of the thirteen states, instead of unanimously, as required by the Articles of Confederation. Perhaps, if the politics was auspicious, the international community would accept a U.N. Charter review conference providing that a new Charter go into effect despite a permanent member veto.[10]

While creating the political will to amend the U.N. Charter would be very difficult, even assuming the problem of the veto could be dealt with, a GPA initiated by way of Charter reform would likely be accepted as the most legitimate.

B. Creation by the United Nations General Assembly as a Subsidiary Organ

Article 22 of the United Nations Charter empowers the General Assembly to "establish such subsidiary organs as it deems necessary for the performance of its functions."[11] The proposal that the General Assembly acting under Article 22 create a parliamentary assembly as a "subsidiary organ" has been suggested on several occasions over the years. For example, Erskine Childers and Brian Urquhart endorsed this approach in their 1994 book, *Renewing the United Nations System.*[12] Recently it has been proposed by the Committee for a Democratic U.N.[13] The idea is attractive in that it provides a way around the cumbersome United Nations Charter amendment process, but it is not without political difficulties of its own.

Whether a parliament can be properly characterized as a *subsidiary organ* of the General Assembly and whether it can be properly deemed *necessary*

---

[10] JOSEPH PRESTON BARATTA, THE POLITICS OF WORLD FEDERATION: UNITED NATIONS, UN REFORM, ATOMIC CONTROL 16 (2004).

[11] U.N. Charter art. 22.

[12] ERSKINE CHILDERS & BRIAN URQUHART, RENEWING THE UNITED NATIONS SYSTEM (1994).

[13] ANDREAS BUMMEL, COMMITTEE FOR A DEMOCRATIC U.N., DEVELOPING INTERNATIONAL DEMOCRACY: FOR A PARLIAMENTARY ASSEMBLY AT THE UNITED NATIONS 75 (2005) (Strategy Paper), http://www.uno-Komitee.de/en/documents/unpa-paper.pdf.

for the performance of its functions[14] is legally questionable in that the parliament would not be answerable to that body. Indeed, the entire rationale for
a parliament is to introduce into global decision making an independent popularly representative body. While the General Assembly has in the past, established autonomous entities such as the United Nations University, none of
its creations have been intended to be an independent source of political authority. The International Court of Justice has opined in the 1987 *United
Nations Administrative Tribunal* advisory opinion that the General Assembly
cannot delegate powers to a subsidiary organ that it does not itself possess or
are not implied as consistent with the overall structure of the Charter.[15] Since
the General Assembly does not have the power to represent directly the citizens of the world, and the United Nations is structured under the Charter as
an interstate organization, opponents of the project could challenge the General Assembly's powers to create a parliament.

Regardless, however, of the General Assembly's actual legal authority to
create a parliament, the United Nations has no institutional mechanism to
prevent a resolute Assembly from acting. Rather, in a political conflict where
more than a few governments will oppose the General Assembly's creation
of a parliament as a perceived threat to their power, legal arguments would
become fodder in the political debate. Of significance in determining whether the parliament's opponents would prevail is whether the decision by the
General Assembly to create a parliament would be regarded as an "important
question" under Article 18 of the Charter requiring a two-thirds as opposed
to majority vote.[16] While Article 18 specifies certain voting matters as important questions,[17] additional unspecified matters are also according to its
terms important questions, but there is a surprising lack of precedent on
which other matters qualify. Specifically for our purposes, as most subsidiary

---

[14] U.N. Charter art. 22 (emphasis added).

[15] The Effects of Awards of Compensation Made by the United Nations Administrative
Tribunal, Advisory Opinion, 1954 I.C.J.47 (July 13), case summary *available at*
http://www.icj-cij.org/docket/index.php?sum=113&code=unac&p1=3&p2=4&case=21&k=d
2&p3=5.

[16] U.N. Charter art. 18, para. 2.

[17] These questions include:

> recommendations with respect to the maintenance of international peace and securi-
> ty, the election of the non-permanent members of the Security Council, the election
> of the members of the Economic and Social Council, the election of members of the
> Trusteeship Council . . . the admission of new Members to the United Nations, the
> suspension of the rights and privileges of membership, the expulsion of Members,
> questions relating to the operation of the trusteeship system, and budgetary ques-
> tions.

Id.

organs have been approved by consensus, the requisite vote required for their establishment is unclear.

Whichever majority is required, however, the overall decision-making structure of the United Nations does not favor the forces of institutional change. Guardians of the status quo have historically enjoyed great success in keeping reform proposals from gaining enough initial traction to appear on the General Assembly's agenda. Most initiatives have quietly died in committees or have otherwise been buried in bureaucracy.

A related problem is that the need to gain the requisite support within the General Assembly for the establishment of a parliament suggests the need for problematic political concessions. For example, presumably responding at least in part to such concerns, the Committee for a Democratic U.N. proposes in its paper that its parliamentary assembly be composed initially of representatives of national parliaments with direct popular elections to occur at an indefinite time in the future[18] and that all U.N. member states could send representatives to the parliament, regardless of whether they come from a legitimately democratically elected parliament.[19]

There is nothing inherently wrong with beginning as a parliament of parliamentarians. In fact, in favor of this approach is the weight of historical example. The European Parliament, the most successful example of the creation of a transnational parliament, began that way in the earliest days of European integration and fulfilled its promise to convert to direct popular election in 1979.[20] Yet, there are dangers in this approach. As has happened in

---

[18] BUMMEL, *supra* note 13, at 78-79.

[19] *Id.* at 90-91.

[20] The founding treaties of what later become the European Union provided that members should be initially appointed to the parliament by their own national parliaments, but that direct elections should occur at a time in the future when the European Council adopts appropriate arrangements. *See* DAVID JUDGE & DAVID EARNSHAW, THE EUROPEAN PARLIAMENT 26-44 (2003). For the most important organic treaty, *see* Treaty Establishing the European Community art. 19, Dec. 24, 2002, 2002 O.J. (C 325) 33, *available at* http://eur-lex.europa.eu/en/treaties/dat/12002E/pdf/12002E_EN.pdf. The European Parliament first submitted a draft convention for direct popular elections in 1960, but the European Council did not adopt it or forward it to the member states. By 1974, the problems caused by representatives having dual mandates (to their own parliaments and to the European Parliament) had become increasingly apparent, and the Parliament submitted a revised draft convention, which was adopted in 1976. *See* SCHELTO PATIJN, EUROPEAN PARLIAMENT, POL. AFF. COMM. RAPPORTEUR, ELECTIONS TO THE EUROPEAN PARLIAMENT BY DIRECT UNIVERSAL SUFFRAGE 11-12 (1974) (Draft Convention with Explanatory Statement), *available at* http://aei.pitt.edu/5184/01/000478_1.pdf; *as adopted,* COUNCIL DECISION AND ACT CONCERNING THE ELECTION OF THE REPRESENTATIVES OF THE EUROPEAN PARLIAMENT BY DIRECT UNIVERSAL SUFFRAGE, (Sept. 20, 1976), *available at* http://www.bundeswahlleiter.de/europawahl2004/downloads/decisionandact.pdf.

other interparliamentary bodies, national parliamentarians may come to feel a sense of ownership in the parliament and be reluctant to promote the evolution toward independent elections. And, every day that elections are extended will delay the growth in the parliament's political influence. Without the public ritual of popular elections to draw publicity and legitimize the parliament, the organization would be unlikely to be noticed. Also, with the national parliamentary representatives' job security dependent upon reelection to their own national parliaments, their day jobs will remain their primary focus. Unlike parliamentarians who are elected specifically to serve in the GPA, national parliamentarians would not see their careers and reputations as tied to building the growth and influence of that organization. Instead, for them it will be primarily a networking forum where issues of common concerns can be discussed with colleagues from other national parliaments.

More troubling is the suggestion that all U.N. member states, regardless of whether they possess democratically elected parliaments, send representatives to the United Nations Parliament. This would undermine the credibility of the organization and compromise its ability to act as an alternative to authoritarianism.

C. Civil Society Organized Elections

The third approach to creating a GPA is for major actors from international civil society to establish a provisional structure for the parliament themselves and to organize and carry out elections. If this approach were followed, the parliament would start as an unofficial body and its empowerment would be reliant exclusively upon its unique claim to a popular mandate described above.

This is the strategy for creating the parliament that my colleague Professor Richard Falk and I first proposed when we began advocating for a GPA.[21] It is also the approach suggested by George Monbiot in his book *A Manifesto for a New World Order.*[22] As we explained in the year 2000 in the *Stanford Journal of International Law:*

> [A] GPA need not be established by a traditional inter-state treaty arrangement. Globalization has generated an emergent global civil society composed of transnational business, labor, media, religious and issue ori-

---

[21] Richard Falk & Andrew Strauss, *On the Creation of a Global Peoples Assembly: Legitimacy and the Power of Popular Sovereignty*, 36 STAN. J. INT'L L. 191 (2000); Andrew Strauss & Richard Falk, *For a Global Peoples' Assembly*, INT'L HERALD TRIB., Nov. 14, 1997

[22] GEORGE MONBIOT, THE AGE OF CONSENT: A MANIFESTO FOR A NEW WORLD ORDER 88-98 (2003).

ented citizen advocacy networks with an expanding independent capacity to initiate and validate a GPA.

. . . .

. . . Uniquely, a GPA would have a claim to authority independent of whether or not it received the formal blessings of the state system.[23]

To begin such a civil society initiating process one might envision a call emanating from a panel of political and moral authority figures, such as former heads of state, Nobel Peace Prize winners and major religious figures. If a critical mass of respectable civil society organizations responded positively to this call, the panel could oversee a series of civil society meetings culminating in a final conference whose purpose would be to adopt a political framework for the parliament's creation. Civil society would then have the task of organizing and holding elections. Presumably, elections would occur in all countries where they were not banned and political conditions allowed for free campaigning.

Needless to say, all of this would be extremely difficult to implement both politically and logistically. Civil society is inchoate and has no preexisting structure for making collective decisions. Putting in place the decision-making process for less ambitious projects such as the World Social Forum has been difficult and contentious,[24] and that project in particular has worked largely because its decentralized nature has kept the need for common decision making to a minimum.[25] Creating out of whole cloth a widely agreed upon decision-making structure, capable of resolving such politically fraught topics as provisional voting formulas and electoral districts would be daunting, even for a skilled panel of authority figures.

The project may become more politically manageable by substituting initiation of the GPA by existing political parties for civil society as a whole. While also lacking a process for making collective political decisions, such parties, numbering far fewer than civil society organizations in general, are likely to be less unwieldy. In addition, they already provide the infrastructure

[23] Falk & Strauss, *supra* note 21, at 194, 206-207 (citation omitted).

[24] For a discussion of the early organization of the Forum, *see* Naomi Klein, *A Fete for the End of The End of History*, THE NATION, Mar. 19, 2001, at 19, 22. ("The organizational structure of the forum was so opaque that it was nearly impossible to figure out how decisions were made or to find ways to question those decisions. There were no open plenaries and no chance to vote on the structure of future events. In the absence of a transparent process, fierce NGO brand wars were waged behind the scenes . . . ."); *see also* Teivo Teivainen, *World Social Forum: What Should It Be when It Grows up?*, OPENDEMOCRACY, July 10, 2003, at 3, http://www.opendemocracy.net/content/articles/PDF/1342.pdf.

[25] The World Social Forum allows civil society organizations to self-organize their own events under the umbrella of the Forum.

for electoral politics and might look favorably on an opportunity to expand their arena. Regardless, however, of which nongovernmental organizing entities were to take the initiative to begin the parliament, the barriers to reaching agreement and acting on that agreement are significant.

Finally, funding would have to be secured to underwrite the cost of the elections and the initiation of the parliament. If the costs of domestic elections and operating expenses of existing parliaments are a guide, the sums would greatly exceed the amounts that have thus far been devoted by the nongovernmental sector to international political initiatives.

D. An Interstate Treaty Process

Finally, a GPA could be established by way of a stand-alone treaty agreed to by whichever internationally progressive countries were willing to be pioneers. Even twenty to thirty economically and geographically diverse countries would be enough to found the parliament. The treaty agreed to by these countries would establish the legal structure for elections to be held within their territories including a voting system and electoral districts. In addition, an operational framework for the parliament, including its mandate and limitations on its powers would be included in the treaty as would a provision for future accession by other countries. Any country could later join the parliament so long as it was willing to meet its obligations under the treaty, the most important of which would be to allow its citizens to vote representatives to the Parliament in free and fair elections.[26]

A stand-alone treaty organization whose membership may not be the same as the United Nations is not a novel concept. Most major international bodies such as the Bretton Woods organizations, the World Trade Organization and the World Health Organization, to name but a few, have been created in this way. Most significant, this approach was used to establish the International Criminal Court, whose membership famously does not include the United States, nor for that matter Russia or China (though Russia is a signatory).[27] In the case of the International Criminal Court, specific treaty provisions align that organization's processes with those of the United Nations. Most significant are terms providing for the Security Council to refer criminal cases to the Court.

---

[26] Professor Falk and I have developed this argument in several places, including *Foreign Affairs* and *The Nation. See* Richard Falk & Andrew Strauss, *Toward Global Parliament,* 80 FOREIGN AFF. 212 (2001); Richard Falk & Andrew Strauss, *Toward a Global Parliament,* THE NATION, Sep. 22, 2003, at 28.

[27] International Criminal Court, Establishment of the Court, http://www.icc-cpi.int/about/ataglance/establishment.html.

Likewise, the GPA treaty could also include provisions defining its initial role *vis a vis* the United Nations, and once established the parliament could enter into a relationship agreement with that body.[28] It would be important to be clear that the parliament, though begun independent of the United Nations, was meant to strengthen, and not replace, that organization. Part of the Parliament's treaty-based responsibilities, for example, could be to weigh in with its own vote on certain specified categories of United Nations General Assembly resolutions. General Assembly resolutions are themselves largely recommendatory, and by insinuating a democratic voice into the process, the resolutions that passed both bodies would be more noticed and deemed more legitimate. Backed by the weight of popular authority over time, perhaps the General Assembly and the GPA could evolve together into a truly bicameral legislative system capable of producing binding legislation.

This approach to creating a GPA by interstate treaty process is the one that Richard Falk and I have come to promote as the most promising. It offers strategic advantages as compared to either of the two proposals for creating the Parliament through the machinery of the United Nations. Even under the second relatively less cumbersome process of the General Assembly voting to create the parliament as a subsidiary organ, a core group of sponsoring countries would have to overcome a formidable combination of bureaucracy, indifference and opposition to gain traction within the United Nations. Under the stand-alone treaty approach, however, power would shift to those countries that are willing to proceed on their own. No one could stop them. And once it became clear that the GPA treaty initiative had left the station, it would likely gain momentum as other less proactive countries would have an incentive to take part rather than be sidelined in the creation of an important new international organization.

---

[28] The U.N. Charter provides that: "The Economic and Social Council may enter into agreements with any" agency "established by intergovernmental agreement and having wide international responsibilities." U.N. Charter art. 63, para. 1; *id.* at art. 57, para. 1. These agreements "defin[e] the terms on which the agency concerned shall be brought into relationship with the United Nations[,]" and "shall be subject to approval by the General Assembly." U.N. Charter art. 63, para. 1. Relationship agreements typically provide for the exchange of information, common facilities and assistance, cooperation in financial and administrative matters, and the mode of debate and agenda setting. *See* U.N. Joint Inspection Unit, *Relationship Agreements Between the United Nations and the Specialized Agencies: Review and Strengthening of Sections Pertaining to the Common System of Salaries, Allowances, and Conditions of Service*, U.N. Doc JIU/REP/93/3 (1993), *available at* http://www.unsystem.org/jiu/data/reports/ 1993/en93_03.pdf. Relationship agreements may also reaffirm the independence of the organization entering into relationship with the United Nations, as did the International Criminal Court in its relationship agreement with the U.N. International Criminal Court, International Cooperation, http://www.icc-cpi.int/about/ataglance/cooperation.html.

Beyond this strategic leveraging of support, countries that are truly supportive of the GPA's democratic mission are likely to create the best, most democratic organization. They would not be forced to make the kinds of antidemocratic concessions that passage by the United Nations might require. Later, if a critical mass of countries were to join the parliament, there might come a time when it would be politically untenable for holdout governments to deny their people the right to vote in the only globally elected body. At that point those governments would not be in a position to compromise the integrity of the organization, but would have to join the GPA on its own democratic terms.

Finally, relative to civil society organizing elections, an interstate treaty process does not suffer from the absence of a decision-making structure that would undermine the ability of nongovernmental organizations to act collectively. States have a long-accepted and highly defined collaborative process for entering into treaty arrangements, including those establishing new international organizations. Also, state sanction for the GPA by way of treaty would confer an additional layer of legitimacy upon the organization, and states have access to the resources to finance the project that civil society lacks.

IV. Conclusion

We all belong to multiple geographical communities, local, provincial, national and international. People who consider themselves committed democrats often take it as a first principle of politics that all of these communities should be organized along democratic lines—all that is except for the international whose democratic failings many commonly overlook. In the age of globalization this exception is anachronistic and dysfunctional. It means that the normal dynamics of parliamentary politics stop at water's edge. In domestic democratic politics parliamentary coalitions are fluid, and while they are affected by party loyalties and discipline, such coalitions transcend geographical boundaries. Parliamentarians, representing voting constituencies, stand with some colleagues on some issues and other colleagues on other issues regardless of from where they hail.

In the international system, on the other hand, all of the citizens of a certain nationality, and/or within a certain geographical area, are part of a permanent coalition frozen into the institutional mold of the state. This permanent coalition in theory, and largely in practice, speaks with one voice in international affairs. It doesn't matter that citizens within the state may find that on certain global issues their political affinities or interests match more closely with the positions taken by other states.

This structure gives those who capture political control of the permanent coalition that is the state the ability to wage war. At their disposal is the capacity to field organized armed forces fueled by nationalist sentiment. Parliamentary coalitions, on the other hand, because of their constantly changing conditional composition are not centrally controlled, and do not engender nationalistic feelings of group identity, nor do they field armies. While violence can break out within the context of a parliamentary system, it is antithetical to, rather than consistent with, the structure of that system, a system which is after all dedicated to the principle of irenic decision making.

In this article I have tried to explore concretely how such a parliamentary system might be introduced into the global arena. With the understanding that what is feasible and practical is highly contingent upon unknown political developments, my goal has been to present what amounts to an historical work-in-progress rather than final conclusions. As David Kennedy has, for example, argued elsewhere in this volume "rupturing historical forces" could dramatically change the political context in which we are operating. What seems highly improbable now may seem suddenly possible, or what seems possible now may seem suddenly highly improbable. The challenge, which I have tried to meet in this article, and the ongoing challenge for all of us who wish to contribute to making the global system more democratic, is to continue to adopt our ideas, strategies and practices to a global political context which is constantly evolving. Then, it can be hoped that, as opportunities present themselves, we will be ready.

3
On How a GPA Would Help Overcome the Dysfunction in the Current Global Law-Making System

Overcoming the Dysfunction of the Bifurcated Global System: The Promise of a Peoples Assembly

by Andrew Strauss

Transnational Law & Contemporary Problems, 1999[*]

I. Introduction

Richard Falk and I have proposed that the time is ripe for global civil society to take the lead and initiate a popularly representative Global Peoples Assembly (GPA).[1] The tremendous growth in the commitment to, and practice of, democracy in domestic settings[2] juxtaposed against globalization's large scale transfer of political decision-making to interna-

---

[*] Reprinted from TRANSNATIONAL LAW & CONTEMPORARY PROBLEMS, Volume 9, No. 2, Fall 1999.

[1] See Richard Falk & Andrew Strauss, *On the Creation of a Global Peoples Assembly: Legitimacy and the Power of Popular Sovereignty*, 36 STAN.J. INT'L L. (forthcoming Summer 2000) [hereinafter On the Creation]. *See also* Richard Falk & Andrew Strauss, *Globalization Needs a Dose of Democracy*, INT'L HERALD TRIB., Oct. 5, 1999, at 8; Andrew Strauss & Richard Falk, *For a Global Peoples Assembly*, INT'L HERALD TRIB., Nov. 14,1997, at 8; Andrew Strauss & Richard Falk, *All That Dough*, PHIL. INQ., Oct. 12, 1997, at 7.

[2] Over the last thirty years there has been a significant global trend toward democratization. It started in southern Europe in the mid-1970s, engulfed Latin America in the 1980s, and expanded to many parts of Asia, the former Soviet Union, Eastern Europe, and Africa in the late 1980s and early 1990s. According to Freedom House, 117 countries-more than sixty percent of the world's states-are now, at least in a qualified sense, democratic. *See* Adrian Karatnycky, *The Comparative Survey of Freedom 1998-1999: A Good Year for Freedom, in* FREEDOM HOUSE: THE ANNUAL SURVEY OF POLITICAL RIGHTS AND CIVIL LIBERTIES 1998-1999 1, 3 (1999). Reporting on and interpreting this trend toward democracy has not surprisingly been a major theme among commentators. Receiving perhaps the most notoriety has been Harvard political scientist Samuel Huntington who, in 1993, introduced the concept of waves of democratic expansion and surveyed the advance of democracy around the world. *See generally* SAMUEL P. HUNTINGTON, THE THIRD WAVE: DEMOCRATIZATION IN THE LATE TWENTIETH CENTURY (1993). At about the same time, a related work by international law professor Thomas Franck received a great deal of attention within the international law community. Franck, the recently retired president of the American Society of International Law, celebrated this trend toward democratization by suggesting that international law was in the process of

tional institutions[3] has made the almost complete lack of democracy at the international level the most glaring anomaly of the global system today.

Because states are unlikely to initiate the democratization of the international order, the task of beginning the drive for the first GPA necessarily falls to civil society.[4] In taking up this cause, civil society could employ various strategies for bringing about such an Assembly. For example, it could estab-

---

evolving a right to democratic governance. *See* Thomas M. Franck, *The Emerging Right to Democratic Governance*, 86 AM. J. INT'L L. 46 (1992). While the global trend toward democratization heralded by Huntington and Franck continues, or certainly has not lost significant ground, scholars have more recently begun to give greater emphasis to the frailties of existing democratic systems and the potential limits on the further expansion of democracy. *See* SAMUEL P. HUNTINGTON, THE CLASH, OF CIVILIZATIONS AND THE REMAKING OF WORLD ORDER (1996) (asserting that democratic expansion is limited by its connection to western cultures); Fareed Zakaria, *The Rise of Illiberal Democracy*, FOREIGN AFF., Nov/Dec. 1997, at 22 (observing that despite electoral validation, many governments maintain authoritarian systems of administration); Thomas Carothers, *Democracy Without Illusions*, FOREIGN AFF., Jan/Feb. 1997, at 85 (discussing the countermovement away from democracy).

[3] Many observers perceive this evolution of power to the international order. For some important works, see SUSAN STRANGE, THE RETREAT OF THE STATE: THE DIFFUSION OF POWER IN THE WORLD ECONOMY (1996) (arguing that state power is giving way to the power of global institutions and multinational corporations); Christoph Schreuer, *The Waning of the Sovereign State: Toward a New Paradigm for International Law?*, 4 EUR. J. INT'L L. 447, 451 (1993) (discussing the increasing independence from their state constituents of such international organizations as the World Health Organization, the International Monetary Fund, and the World Trade Organization); ROBERT J. HOLTON, GLOBALIZATION AND THE NATION-STATE 50-79 (1998) (noting the evolution towards transnational networks and regulatory arrangements). *See also* GLOBALIZATION AND GLOBAL GOVERNANCE, Part I (Raimo Vayrynen, ed.) (1999) (examining the need for global regulatory institutions).

[4] Now surfacing are several preliminary though important citizen efforts to create an institution similar to what we are calling the Global Peoples Assembly. Out of Perugia, Italy, an initiative called the Assembly of the United Nations of Peoples has attempted to organize civil society organizations into a quasi-representative assembly. Last fall with civil society organizations from 100 countries in attendance, it had its third assembly. Similarly significant is the Global Peoples Assembly Movement, which had its inaugural meeting in Samoa in April. Like the Perugia initiative, this movement's purpose is to create an ongoing institutional structure that would allow the global citizenry to have an effective voice in global governance. The well regarded civil society organization Earth Action has, as part of its "Call for a Safer World," endorsed a directly elected peoples assembly and is launching a major effort to begin an organizational drive to create a treaty body this year. In addition, a major discussion is currently going on within the World Federalist Association on whether it should return to its roots and itself participate in an organizing drive for a Global Peoples Assembly. Perhaps most important, and linking all of these initiatives, is the Millennium NGO Forum. At the invitation of the United Nations Secretary General, representatives of hundreds of civil society organizations will convene this May at the United Nations Headquarters in New York. One of the primary stated goals of the forum is "to create an organizational structure whereby peoples of the world can participate effectively in global decision-making." The forum's outcome will be reported on by the Secretary General to a special millennial assembly of states examining the future architecture of the global system of governance.

lish an embryonic assembly composed of representatives of civil society organizations with the goal that this body evolve into a popularly elected assembly. Alternatively, a very hopeful approach might be to enlist a relatively small core of like-minded states to create a treaty based electoral assembly to which other countries could over time be persuaded to join. Professor Falk and I have suggested that with the help of any willing states civil society could itself organize elections and establish the GPA. Quite clearly this strategy would entail overcoming formidable challenges. Civil society would have to come up with an institutional mechanism that would be broadly accepted for establishing electoral districts as well as campaign finance and other election rules. The fact that some of the world's more authoritarian governments would almost certainly not allow elections to occur in their countries would have to be dealt with. Until sufficient pressure could be brought to bear, probably after the Assembly was well established, citizens of these countries would most likely have to go unrepresented. In other seemingly more receptive countries, civil society would have to guard against attempts to manipulate elections. Once the Assembly was constituted, meeting facilities, translation services, office support personnel and other staff would have to be provided. All of this would call upon civil society to do more than it has ever done before. Nevertheless, in light of civil society's recent successes in initiating global reform, Professor Falk and I have suggested that it has the potential (we hope with the help of receptive states) to carry out such a project.[5]

If a citizen-founded Global Peoples Assembly were in fact to become a 21st century reality, it would not, of course, immediately transform global governance. As Section III of this article will explain, it would rather establish an institutional structure that would have the potential over time to grow into a legislative body with accepted law-making powers. If such an Assembly were eventually to evolve into one with a significant legislative role, global governance would likely be improved in several respects. In this article I will, however, only explore how the Assembly could help overcome the most fundamental institutional dysfunctions of the current international system. To function effectively, any community legal system must be able to prescribe norms of behavior for the community.[6] Once such norms are prescribed, the system should be structured so as not to discourage internal compliance,[7] and finally, the system should have the capacity to enforce its

---

[5] *See* Falk & Strauss, *On the Creation, supra* note 1.

[6] *See* DONALD BLACK, THE BEHAVIOR OF LAW 105-121 (1976).

[7] *Id.* at 61-83.

norms.[8] The international system in significant respects fails these most basic tests of functionality.

At present, to a great extent in both theory and practice, elites who control the mechanisms of state power can choose for their states not to be bound by those international laws they do not wish it to be bound by and even once bound, such elites maintain the internal ability to organize their citizenry to facilitate state non-compliance with the law when they deem it expedient. In response the international system is by and large unable to take effective remedial action to enforce its norms against those individuals who are actually responsible for causing international law violations. What all this means as a practical matter is that those who control state power can on behalf of their states often *de jure* or *de facto* opt out of community law.

In this article, I will trace the root of the dysfunction to what I call the bifurcated global system, by which I mean the rigid division of the global system into two distinct polities, the domestic and the international. In place of what, with the help of a directly representative assembly, could be an unmediated democratic connection between the global citizenry and the international order, states at present intermediate this relationship. Not only does this result in citizen access to international law-making power being channeled through states, but correspondingly, with rare exception, states, at the expense of the international order, maintain a captive monopoly on the legal obligations of citizens. Instead of a seamless democratic global legal system where no one is above the law, what results is a dysfunctional system which allows states to stand in contravention of the law, a system, to paraphrase John Adams, of states and not of laws.[9] Tracing the origins of this dysfunction of the international system and understanding how a cure might be found in the promise of a Peoples Assembly is the subject of what follows.

II. The Dysfunction of the Bifurcated Global System and the Promise of the Global Peoples Assembly

A. State Intermediation and the Bifurcated Global System

The global system as presently structured relies on states to be intermediaries between citizens and the international system. This means that the planet's six billion citizens are not directly involved in creating international law, and the international legal order, likewise, does not directly command their compliance with its laws. Because law, domestic or international, can only affect

---

[8] *Id.* at 86.

[9] *See* John Adams, 2 PAPERS OF JOHN ADAMS 314 (Rubert J. Taylor ed., 1977).

the social order to the extent it influences the behavior of real human beings, both the process of creating and complying with international law necessarily must take place in two steps. With some recent qualification,[10] if citizens wish to influence the creation of international law, they must petition their own government, which can, if it chooses, respond favorably to their appeal and work toward the creation of such law.

Once a state has agreed to a law, it must command real persons to do, or to refrain from doing, whatever is necessary to bring itself into compliance with the law. If states wish to establish an effective international regime banning the production of ozone-harming chemicals, for example, they must first enter into an agreement making such chemicals illegal under international law and then each must preclude the real individuals within its jurisdiction from producing such chemicals. Sometimes the demands might be upon individuals who are acting as agents of the states either by direct employment or by commercial contract. For example, if two countries wish mutually to reduce their stockpiles of nuclear weapons, they must first enter into an international legal agreement among themselves to do so, and then each country must make the requisite demands upon the individuals involved in their respective defense establishments so that compliance with the agreement is achieved. Alternatively, if these two countries wish to agree to create a military alliance with each other, they would enter into an international legal agreement and then command those individuals in their respective defense establishments to act in such a way that the terms of the alliance were met.

B. The Peoples Assembly and the Dysfunction of the Bifurcated International Law-Making System

This bifurcated system, requiring states to intermediate between citizens and the international order, is quite dysfunctional. It charges states with setting

---

[10] Non-governmental organizations, often now referred to as civil society organizations, are increasingly attempting to bypass the nation-state and participate directly at the international level. For further discussion, see generally Peter J. Spiro, *New Global Communities. Nongovernmental Organizations in International Decision-Making Institutions*, WASH.Q., Winter 1995, at 45; Dianne Otto, *Nongovernmental Organizations in the United Nations System: The Emerging Role of International Civil Society*, 18 HUM. RTS.Q. 107 (1996); Dinah Shelton, *The Participation of Non- Governmental Organizations in International Judicial Proceedings*, 88 AM J. INTL L. 611 (1994). While international organizations have been reluctantly and tentatively opening their doors to citizen organizations, within the logic of the present international system, constituted as a community of states rather than citizens, there is no clear structural role for citizen organizations to play.

the rules for the international system,[11] including the foundational rules which determine the extent to which states are imposed upon to comply with community norms.[12] States have used this power to further their ability to maintain absolute discretion over the extent of their participation in the system. States, or more precisely those who control state power, have done this by adopting a rule that allows each state to decide for itself which international laws it wishes to be bound by.[13] Thus, for example, a state is under no

[11] The dominant view is that under the bifurcated international legal system, international law emanates from the general consent of states. *See*, e.g., Louis Henkin, *International Law: Politics, Values and Functions*, 216 RECUEIL DES COURS 46 (1989) (proclaiming that "[i]nter-state law is made, or recognized, or accepted, by the 'will' of States. Nothing becomes law for the international system from any other source"). This view, while generally endorsed by those who control the mechanisms of state power, has been contested. *See* MYRES MCDOUGLAS ET AL., STUDIES IN WORLD PUBLIC ORDER (1960); RICHARD FALK, LAW IN AN EMERGING GLOBAL VILLAGE: A POST-WESTPHALIAN PERSPECTIVE (1998).

[12] *See* THOMAS FRANCK, FAIRNESS IN INTERNATIONAL LAW AND INSTITUTIONS 11-13 (1995) (explaining why the international community of states as a "rule community" has created its own process for making and applying rules and resolving disputes about their meaning).

[13] Generally speaking, states have accepted the view that they are only bound by those international laws to which they agree, either explicitly by treaty or impliedly by customary practice. This statement, however, draws upon a long tradition of discussion and debate and is subject to some qualification. First, in regard to customary international law, states have traditionally maintained that they are only bound by customary laws that they accept. The generally accepted requirement, however, for manifesting a lack of consent ("the persistent objector rule") demands that states declare their intention not to be bound by a customary law at the time of the formation of the law. *See* Luigi Condorelli, *Custom, in* INTERNATIONAL LAW: ACHIEVEMENTS AND PROSPECTS 179, 205 (Mohammed Bedjaoui ed., 1991). For a variety of reasons, this is sufficiently unlikely to happen that, practically speaking, most of the time states are considered bound by customary international law. Second, and more fundamentally, there has been an increasing, though seldom articulated, trend toward viewing customary international law as binding on even those states that have stated a desire not to be bound based upon a subtle shift to the notion that a general consensus among states binds all states. For further discussion, see J. Patrick Kelly, *The Twilight of Customary International Law*, 40 VA J. INTL L. 449 (2000). Most states, however, do not seem to accept this view.

Treaties are to an ever-greater extent replacing customary international law as the primary source of legal obligations on states. Whether a state wishes to adhere to a treaty is, under the state created rules of the international system, always voluntary. Here, however, some qualification is also in order. While a state always has the legal option to resist a treaty, as a practical matter, in certain cases diplomatic pressure may be such that a state actually has little option but to accept a treaty. For a discussion of this problem, see Robert Keohane, *Reciprocity in International Relations*, 40 INT'L ORG. 1 (1986).

The final qualification to the absolute ability of states to opt out of international law is that most states seem to accept that in certain very limited cases fundamental international norms can apply to states that do not in any way consent to be bound by them. They hold that, regardless of consent, states can be bound by natural law (some concept of absolute right) not to engage in certain basic wrongs. *See*, e.g., Vienna Convention on the Law of Treaties, May

legal obligation to agree to be bound by community laws that are necessary to the safety and well being of the international community. Overtures by the international community to get states to join in pollution control, weapons elimination, or other regimes that are vital to the interests of the global community can all be rejected legally.

The implications of this for the international system are significant. Just as would-be bank robbers are among the least likely citizens to agree voluntarily to subject themselves to domestic laws proscribing bank robbery, those states contemplating action contrary to an international law are the least likely to assent to that law. For example, India's 1995 refusal to agree to be bound by the nuclear non-proliferation treaty[14] might well have foretold its intention to become the newest avowed nuclear power.[15] When, after three years, this came to pass, it precipitated a dangerous arms race in South Asia.[16]

The problems caused by the ability of states to opt out of community norms, however, go beyond the system's inability to extend the law to a limited number of recalcitrant countries. Because treaties are often ineffective without the participation of certain, and sometimes most, countries, a defiant

---

23, 1969, art. 53, 1155 U.N.T.S. 331, 334 (1969) (providing that a treaty is "void if ... it conflicts with a peremptory norm of general international law"). Peremptory norms are probably inclusive of core human rights standards. *See* LAURI HANNIKAINEN, PEREMPTORY NORMS (JUS COGENS) IN INTERNATIONAL LAW: HISTORICAL DEVELOPMENT, CRITERIA, PRESENT STATUS, Part III (1988).

[14] Treaty on the Non-Proliferation of Nuclear Weapons, *opened for signature* July 1, 1968, 21 U.S.T. 483, 729 U.N.T.S. 161. In 1995, India refused to agree to an extension of the nuclear nonproliferation treaty. Almost all of the non-nuclear weapons states have joined the nonproliferation treaty which requires them to remain nonnuclear. The five nuclear weapons states party to the treaty have accepted the treaty's requirements that they eventually eliminate their nuclear weapons. Only India, Pakistan, Israel, and Cuba have not joined the global regime. Cuba is not thought to have nuclear weapons. Israel does, but has not declared itself to be a nuclear weapons state. Only India and Pakistan have tested nuclear weapons and declared their nuclear status

[15] By using India's failure to adhere to the extension of the nuclear nonproliferation treaty, I do not mean to imply that India's arguments against the treaty were without validity. India argued that there had, in fact, not been a good faith effort on the part of the nuclear powers to meet the treaty's nuclear disarmament requirement and that the treaty, therefore, acted to legitimize a discriminatory state of affairs. India was particularly concerned that China, with whom it shares a border and a history of tension, continued to retain such weapons.

[16] In May of 1998, India conducted nuclear weapons tests and declared itself to be a nuclear weapons state. Fifteen days later, rival Pakistan, claiming it had no other strategic choice but to follow India's lead, conducted its own series of tests and also declared itself a nuclear power. For further discussion of the implications of this for regional and global security, see Strobe Talbott, *Dealing with the Bomb In South Asia,* FOREIGN AFF., March/April, 1999, at 110; *The Most Dangerous Place on Earth?,* ECONOMIST, May 22, 1999, at 5.

minority can effectively veto the introduction of treaties that are in the vital interest of the world community. This is the practical, but reversible, effect of the United States' refusal thus far to ratify the Kyoto Protocol,[17] the treaty that establishes binding limits on greenhouse gases.[18] The failure to achieve this treaty could have major adverse implication for the global climate,[19] but going forward with a treaty makes no sense without participation by most countries and especially the United States, which accounts for over twenty-five percent of the world's greenhouse gases.[20]

   To only focus on obvious failures — the refusal of individual states to accede to international norms, or the potential for a minority of states to effec-

---

[17] Kyoto Protocol to the Framework Convention on Climate Change, Report of the Conference Parties, 3rd sess., Agenda Item 5, U.N. Doc. FCCC/CP/1997/L.7/Add. 1 (1997), *reprinted in* 37 I.L.M. 22 (1998) [hereinafter Kyoto Protocol].

   My assertion that the United States is acting to "veto" the Kyoto Protocol needs some explanation. In December 1997, the United States joined 150 other countries in Kyoto, Japan, under the U.N. Framework Convention on Climate Change to finalize negotiations over an agreement to provide for mandatory reductions in greenhouse gases. After difficult negotiations, a final agreement was reached which provided a comprehensive plan to reduce greenhouse gas emissions between 2008 and 2012. Developed countries are required by the Protocol to reduce emissions significantly below 1990 levels; however, the requirements for developing countries are far less stringent. President Clinton signed the protocol. Strong opposition to the treaty, however, continues within the United States Senate because of stated concerns over the lack of "meaningful participation" by developing countries and the impact of the treaty on the U.S. economy. The President has, therefore, not yet presented the treaty to that body for ratification, and has been attempting to negotiate with developing countries so that they might assume greater responsibility for the reduction of greenhouse gases.

   While most other countries have also not yet ratified the convention, their reluctance has not principally been opposition to the treaty but a desire to see what would happen in the United States. While it is certainly too soon to say that the United States has permanently defeated the treaty, and an ultimately stronger treaty with increased developing country participation might emerge, it is not clear that the U.S. Senate, unless ideologically reconstituted, will ever ratify the treaty. Certainly the United States is causing a delay in the full implementation of measures necessary to deal with the problem of global warming.

[18] For an overview of the Kyoto Protocol, see Claire Breidenich et al., *The Kyoto Protocol to the United Nations Framework Convention on Climate Change*, 92 A.J.I.L. 315 (1998).

[19] *See generally* WORLD METEOROLOGICAL ORGANIZATION/UNITED NATIONS ENVIRONMENT PROGRAMME INTERGOVERNMENTAL PANEL ON CLIMATE CHANGE, CLIMATE CHANGE: THE IPCC SCIENTIFC ASSESSMENT (J. T. Houghton et al. eds., 1990). For a technical examination of the problem of global warming, see S. GEORGE PHILANDER, IS THE TEMPERATURE RISING? (1998). For a less technical account, see ED AYRES, GOD'S LAST OFFER: NEGOTIATING FOR A SUSTAINABLE FUTURE (1999).

[20] *See* Mike Dunn, *Conference to Affect State: Nations Will Gather to Discuss Global Warming,* SOLUTIONS, THE ADVOCATE (Baton Rouge, La.), Nov. 30, 1997, at 1B. Realizing this, the Protocol's drafters provided that the Protocol needs to be ratified by 55 parties representing at least 55% of global carbon emissions in order to come into force. *See* Kyoto Protocol, *supra* note 17, art. 24.

tively veto a treaty — is still to understate the dysfunctionality of the present system. Given the intransigence that so many states exhibit in so many different types of negotiations, when broad based agreements capable of securing general adherence are reached, it is often done at the cost of sacrificing a treaty's true effectiveness. So riddled with ambiguities and exceptions or reservations, for example, are conventions to protect women, such as the Convention on the Elimination of Discrimination Against Women[21] and the Convention Concerning Equal Remuneration,[22] that these lowest common denominator instruments are severely compromised in their effectiveness.

An effective Global Peoples Assembly would go a long way towards remedying this dysfunction of the international law-making system. The delegates to the Global Peoples Assembly, unlike national elites who control the mechanisms of state power, would not have an interest in promoting unfettered discretion by states over whether or not they are to be bound by international legal obligations. Therefore, as distinct from the current international legislative system controlled exclusively by states, a Global Peoples Assembly, committed to the rule of law, would be much more predisposed to proclaim a functional system of binding law that states could not opt out of. Supporting the right of the Assembly to issue such a proclamation and legitimizing it in the public mind would be the fundamental democratic principle that the ultimate power to create law flows from the consent of the governed, and that states do not have the discretion to opt out of law that is authorized by the higher authority of the citizenry.[23] If the GPA were over time to become an accepted part of the international law-making system, possibly integrated into an expanded United Nations, such a proclamation could become accepted as a fundamental legal precept of the international system. After all, what appears so obvious in a domestic setting — that no functional legal system can give its subjects the individual discretion to opt out of laws they

---

[21] Convention on the Elimination of All Forms of Discrimination Against Women, *opened for signature*, Mar. 1, 1980, G.A. Res. 341180, U.N. GAOR, 34th Sess., Supp. No. 46, at 193, U.N. Doc. A/34/46 (1979), 1249 U.N.T.S. 13, *reprinted in* 19 I.L.M. 33 (1980) (entered into force on Sept. 3, 1981). This convention has had a particularly large number of reservations. *See* William Schabas, *Reservations to the Convention on the Elimination of all Forms of Discrimination Against Women and the Convention on the Rights of the Child*, 3 WM. & MARY J. OF WOMEN & L. 79(1997).

[22] Convention Concerning Equal Remuneration for Men and Women Workers for Work of Equal Value (ILO Convention No. 100), June 29, 1951, 165 U.N.T.S. 303. The language of the convention is so hedged that states are not bound to much of anything. *See especially* art. 2

[23] Of course, this is not to imply that a claimed right of the Assembly to impose law should or would be without limitations. Such a claimed power of the Assembly to create or participate in creating binding law would only be likely to be accepted to the extent that it was constitutionally limited to matters within the proper scope of international decision-making. For further discussion, see *infra* § IV.

do not like — should, once officially proclaimed, come to appear obvious in
the international setting as well.

C. The Peoples Assembly and the Dysfunction of the Bifurcated International Law - Compliance System

Once a state consents to be bound by an international law, it is obliged to
follow that law. But here too, the system remains dysfunctional. Though
states at this point no longer have the legal right not to comply with community norms, the bifurcated system is specifically designed to give them the
practical ability to do just that. Because states (with one significant emerging
exception)[24] continue to be the sole source of law applicable to citizens, they
maintain the ability to mobilize their citizens to directly contravene their
international legal obligations.

Securing institutional compliance with law is different from securing the
compliance of individuals because institutional action which is intended to
be either in conformity or not in conformity with the law is itself largely

---

[24] Following World War II, the notion that at least certain international human rights and
humanitarian norms apply not just to states, but also to individuals, began to gain acceptance.
The watershed events were the Nuremberg and Tokyo trials and convictions of individuals for
war-time violations of international humanitarian and human rights law. *See, e.g.,* DEP'T. OF
STATE, Pub. No. 2420, TRIAL OF WAR CRIMINALS' INDICTMENT 23 (1945); Agreement for the
Prosecution and Punishment of the Major War Criminals of the European Axis, Aug. 8, 1945,
59 STAT. 1544, E.A.S. 472; *International Military Tribunal (Nuremberg) Judgment and
Sentences*, 41 AM. J. INT'LL. 172 (1947). In the last several years, the principle that certain
international humanitarian and human rights laws apply directly to individuals has come
closer to becoming an enforceable reality. For further discussion of the need to disintermedi-
ate enforcement generally, see *infra* § II D.) First, in 1993, the United Nations Security Council established an International Criminal Court at The Hague to try Balkan human rights and
war crimes suspects. The Security Council followed in 1994 by establishing an affiliated court
in Arusha to try Rwandan human rights and war crimes suspects. Both courts are presently in
operation. *See generally* Louise Arbour, *History and Future of the International Criminal
Tribunals for the Former Yugoslavia and Rwanda*, 13 AM. U. INT'L L. REV. 1495 (1998).
Later, in the summer of 1997 in Rome a United Nations conference of states approved a draft
convention to establish for the first time a permanent international criminal court. Upon ratification by 60 countries the convention will go into effect. *See generally* Mahnoush Arsanjani,
*Developments in International Criminal Law: The Rome Statute of the International Criminal
Court*, 93 AM. J. INT'L L. 22 (1999). Because of strong lobbying by the international human
rights community (*see* Falk & Strauss, *On the Creation, supra* note 1) and the horror and
significantly destabilizing effects of the acts proscribed, the international community (with
some major holdouts (*see id.*)) is in the process of providing an institutional mechanism for
certain international criminal laws to be applied directly to citizens. Countries are, however,
for the reasons discussed in this article, very unlikely to expand the scope of the applicability
of international law to individuals beyond this narrow subject area.

organized and coordinated through rules.[25] States wishing to maintain centralized control over their compliance with international law must, therefore, come up with some legal mechanism for coordinating the actions of their citizens in the furtherance of directing either compliance or noncompliance with international law. There are different doctrinal approaches used by states to accomplish this. The United States follows a particular application of what is called dualism which maintains — consistent with the bifurcation of the international system — that international law and domestic law are two totally distinct systems of law.[26] International law applies between states and domestic law applies within states. The American approach, however, provides that international law is automatically incorporated into U.S. federal law so that it becomes the law of the United States.[27] This means that, in the normal course, when political authorities desire for international law to be followed, they simply allow for such incorporation. However, since United States domestic law holds that a later-in- time federal law takes precedence over an earlier-in-time federal law,[28] by passing legislation[29] inconsistent

[25] *See* Abram Chayes, *An Inquiry Into the Working of Arms Control Agreements*, 85 HARV. L. REV. 905, 968 (1972); FRED IKLE, HOW NATIONS NEGOTIATE 8 (1964). Cf. Roger Fisher, *Bringing Law to Bear on Governments*, 74 HARV. L. REV. 1130 (1961).

[26] Consistent with the bifurcation of the global system, both the domestic and international orders have their own separate bodies of law and each has its own distinctive law-applying institutions. International law is created by states through procedures that manifest their consent either explicitly through treaties or impliedly through customary practice. See *supra* note 13. Domestic law, on the other hand, comes exclusively from the domestic system. In the United States, Congress, state legislatures, administrative agencies, and courts, to name a few, are all law-making institutions. International legal problems are largely resolved through international dispute resolution mechanisms, such as diplomacy, international arbitration, or the International Court of Justice, while domestic legal problems are thought to be largely resolved in national legal institutions, most notably domestic courts. *See* J. G. Starke, *Monism and Dualism in the Theory of International Law*, 17 BRIT. Y. B. INT'L L. 66, 68 (1936); Josef L. Kunz, *The "Vienna School" and International Law*, 11 N.Y.U.L. REV. 370, 399 (1934); HANS KELSEN, PRINCIPLES OF INTERNATIONAL LAW 446-47 (Robert W. Tucker ed., 2d ed. 1966).

[27] *See* The Paquete Habana, 175 U.S. 677, 700 (1900) ("[I]nternational law is part of our law, and must be ascertained and administered by the courts of justice of appropriate jurisdiction as often as questions of right depending upon it are duly presented for their determination"). Not all international law, however, is incorporated into United States law. Only what are called "self-executing" treaties and other "self-executing" international agreements, as well as customary international law that is appropriate for application by domestic courts, are incorporated. All other international law is limited to application in international fora. *See* RESTATEMENT (THIRD) OF THE FOREIGN RELATIONS LAW OF THE UNITED STATES § 111 (3) & cmt. c (1987).

[28] The Supreme Court has interpreted the Supremacy Clause of the United States Constitution (U.S. CONST. art. VI, § 2) to support the equal status of self-executing treaties and statutes and consequently has endorsed application of the later-in-time rule. *See* Whitney v. Rob-

with a previous international legal obligation, the United States political branches can effectively direct those subject to its jurisdiction to violate international law.

Some other countries give internal precedence, to a greater or lesser extent, to international law over their domestic law. Those that give precedence as a general rule to international law are referred to as "monist." All countries are ultimately "dualist," however, in the sense that as a result of the bifurcated international system, their own governing institutions (often their constitutions) are the ultimate determiners of which law will be given precedence. As long as it remains the case that states maintain ultimate control over their own internal compliance with international law, they will have the option of organizing their citizens so as to promote the violation of international law.

In a legal system organized to maximize law-compliance, such an option would not exist. For example, within the domestic realm, corporations are somewhat analogous to states in the international realm in that each entity is self-governed through a system of internal rules and yet operates with the context of a larger community legal system. In all countries, however, the larger community legal system attempts to secure the legal obedience of these corporate entities by not only imposing compliance obligations upon

---

ertson, 124 U.S. 190, 194 (1887) ("By the Constitution, a treaty is placed on the same footing, and made of like obligation, with an act of legislation. Both are declared by that instrument to be the supreme law of the land, and no superior efficacy is given to either over the other. When the two relate to the same subject, the courts will always endeavor to construe them so as to give effect to both, if that can be done without violating the language of either; but if the two are inconsistent, the one last in date will control the other . . .").

While the Constitution does not directly address the matter, and there is no case law directly on point, it is most consistent with the overall framework for the American application of international law described in the text that the later-in-time rule should also apply to customary international law. *See* RESTATEMENT (THIRD) OF THE FOREIGN RELATIONS LAW OF THE UNITED STATES § 115 reporter's note 4 ("arguably later customary law should be given effect as law of the United States, even in the face of earlier law"); *see also* Louis Henkin, *International Law as Law in the United States*, 82 MICH.L. REV. 1555, 1562-64 (1984) (customary international law should be given authority equal to United States federal law).

[29] The extent to which the President acting on his own authority can within this American compliance structure command citizens to act contrary to international law remains unclear. Authority for him to do so has been read into the Supreme Court's dictum that courts will give effect to international law "where there is no treaty, and no *controlling executive* or legislative act or judicial decision." The Paquete Habana, 175 U.S. 677, 700 (1900) (emphasis added). *But see* Louis Henkin, *The Constitution and United States Sovereignty: A Century of Chinese Exclusion and Its Progeny*, 100 HARV.L. REV. 853,879 (1987) "Unlike Congress, the President has no general authority to make law that might compete with international law as law of the United States. The President's duty is to 'take Care that the Laws be faithfully executed,' a duty that applies to international law as well as to other law of the land." *Id.*

them but also upon those employees who work within their structures.[30] Indeed, we would regard it as quite illogical for the state to limit intentionally the effectiveness of its ability to secure compliance with its mandates.

A Global Peoples Assembly would very likely improve the international compliance system by extending the obligation to obey international law directly to individuals. Unlike the national governing elites who, acting as agents of their states, have colluded to maintain an international system whereby their claims on citizen legal obligations are uncontested, a GPA dedicated to an effective system of law would be inclined to proclaim transnational law to be binding on citizens. Correspondingly citizens who directly participate in the international system would likely come to accept and even expect that the law their Assembly creates would apply directly to themselves.

To the extent such law actually comes to be generally accepted by citizens as applicable, in what could be labeled "compliance from the inside out," states would lose the ability to mobilize their citizens in contravention of transnational law. The old system would be turned on its head. States would no longer, as under the bifurcated system, be the vehicle that must be relied upon to bring their citizens into compliance with international legal obligations. Instead, because states are in effect the sum of their citizens, citizen compliance with international legal obligations would necessarily result in state compliance as well.

D. The Peoples Assembly and the Dysfunction of the Bifurcated International Law-Enforcement System

Not only does the bifurcated system facilitate the organizational ability of states to break international law in the first instance, but it does not allow for

---

[30] In addition to the fact that in every domestic legal system individuals are always personally obligated to follow the law, as agents of a corporation or otherwise, the American legal system, for example, has developed several additional ways of enhancing the power of the external law to encourage compliance by real persons operating within the corporate structure. For example, racketeering laws can be used to impose extra severe penalties for committing a crime on behalf of an organization. In addition, managers can be held strictly liable under, for example, certain federal environmental laws for the workplace violation of those laws by subordinate employees. *See generally* Joseph G. Block & Nancy A. Voison, *The Responsible Corporate Officer Doctrine - Can You Go to Jail for What You Don't Know?*, 22 ENVT'L. L. 1347 (1992). This principle of strict liability for corporate managers was upheld by the U.S. Supreme Court in United States v. Park, 421 U.S. 658 (1975), and United States v. Dotterweich, 320 U.S. 277 (1943). Likewise, the American system allows shareholders to sue directors derivatively to enjoin action in, or recover damages for, violation of law. *See* Miller v. AT&T, 507 F.2d 759 (3d Cir. 1974) (holding that the business judgment rule cannot insulate directors from liability for criminal acts).

an effective remedial system of enforcement. Under the bifurcated system, states rather than individuals are the exclusive subjects of international law, so enforcement is with rare exception[31] both against states and by states.

Enforcement against states takes several forms. The most far-reaching enforcement powers are given to the United Nations Security Council under Chapter VII of the United Nations Charter.[32] Under Chapter VII, the Security Council has a right either to impose economic sanctions[33] or use military force[34] against a nation whenever it determines "the existence of any threat to the peace, breach of the peace, or act of aggression."[35] More narrowly focused treaty organizations also have certain enforcement powers. For example, the World Trade Organization can authorize members to introduce retaliatory trade restrictions against other states that have violated one of the agreements enforceable by the WTO such as the General Agreement on Tariffs and Trade.[36]

Other organizations, such as the International Labor Organization (ILO), rely on criticizing the practices of countries, or what is sometimes referred to as the "mobilization of shame," in attempts to enforce their norms.[37] What

---

[31] *See supra* note 24.

[32] U.N.CHARTER, arts. 31-51.

[33] *Id.* art. 41.

[34] *Id.* art. 42.

[35] *Id.* art. 39. Article 39 reads in its entirety as follows:

The Security Council shall determine the existence of any threat to the peace, breach of the peace, or act of aggression and shall make recommendations, or decide what measures shall be taken in accordance with Articles 41 [relating to economic sanctions and severance of diplomatic relations] and 42 [relating to military force], to maintain or restore international peace and security.

[36] If the WTO's dispute resolution system determines a party to be in violation of an agreement governed by the WTO and that party refuses to comply with that determination or pay compensation to the aggrieved party, then the aggrieved party "may request authorization from the [Dispute Settlement Body] to suspend the application to the Member concerned of concessions or other obligations under the covered agreements." Understanding on Rules and Procedures Governing the Settlement of Disputes, art. 22, para. 2, Apr. 15, 1994, Marrakesh Agreement Establishing the World Trade Organization, Annex 2, 33 I.L.M. 1125, at 1226.

[37] See Constitution of the International Labor Organization, June 28, 1919, arts. 3 & 4, 49 Stat. 2712, revised through Oct. 9, 1946, 62 Stat. 3485, T.I.A.S. No. 1868, 15 U.N.T.S. 35, amended, June 25,1953,7 U.S.T. 245, T.I.A.S. No. 3500, 191 U.N.T.S. 143, amended, June 22, 1962, 14 U.S.T. 1039, T.I.A.S. NO. 5401, 466 U.N.T.S. 323, amended, June 22, 1972, 25 U.S.T. 3253, T.I.A.S. No. 7987. The main elements of the ILO supervisory machinery which allows for such critical oversight include the following: (1) a reporting system which requires for states that have ratified conventions to provide regular reports on their implementation (arts. 19, 22); (2) "special constitutional procedures" allowing associations of workers or employers (art. 24) or other states (art. 26) to file complaints against states for failing to implement ratified conventions, and procedures for investigating and reporting on such com-

these international enforcement practices all have in common, however, is that the state itself is the unit upon which enforcement is targeted.

Such enforcement against an artificial person as complex and diffuse as a state is both ineffective and violative of basic human rights. Elites who control state power and can cause international law violations often personally benefit or are responsive to those who personally benefit when their country violates international law. Whether they or those to whom they are beholden actually benefit, those elites are seldom forced to face personally the consequences of their actions. If they had from powerful states, their ability to mobilize the machinery of their own state against international enforcers can remove the possibility that any enforcement action will be taken or, if taken, that it can have any significant effect on their state. Those individuals from less powerful countries who cause violations usually fare as well. While their countries may be the target of international enforcement action, with their social power comes the capacity to insulate themselves from the effects of such action. This seems particularly true in the more authoritarian countries where leaders can successfully insulate themselves from the effects of negative public opinion. Perversely, those who have little responsibility for engineering violations of international law often bear the brunt of enforcement action. Obviously, this significantly limits the deterrent effect of the international enforcement system.[38] In addition, collective punishment is contrary to accepted notions of human rights.[39] Even for those who, inconsistently with

plaints (arts. 27-30) and (3) the ILO's procedures for freedom of association which provides for the investigation of interferences with the rights of workers to organize (arts.24-30). For further discussion, see generally Virginia A. Leary, *Lessons from the Experience of the International Labour Organization, in* THE UNITED NATIONS AND HUMAN RIGHTS: A CRITICAL APPRAISAL 580 (Philip Alston ed., 1992); NICOLAS VALTICOS & GERALDO VON POTOBSKY, INTERNATIONAL LABOUR LAW (2d ed. 1995). It should be noted that, germane to our discussion, the ILO is unique among international organizations in that citizen groups, specifically labor unions and industry associations, are along with states officially represented in what is referred to as its tripartite structure. Human rights treaties such as the International Covenant on Civil and Political Rights, Dec. 16, 1966, G.A. Res. 2200, U.N. GAOR, 21st Sess., Supp. No. 16, at 52, U.N.Doc. A/6316, 6 I.L.M. 368, also employ a similar type of supervisory system as a way of "mobilizing shame" to secure enforcement. *See generally* Anne Fagan Ginger, *The Energizing Effect of Enforcing Human Rights Treaty*, 42 DEPAUL L. REV. 1341 (1993).

[38] Trade sanctions authorized by the World Trade Organization, see supra note 36, are actually well designed to provide a measure of deterrence. The aggrieved party, who may decide against which commercial sector to impose trade restrictions, is most likely to choose one with significant domestic political clout. That way the target government will be placed under maximum internal pressure to adhere to its trade obligations.

[39] *See generally* COUNCIL ON FOREIGN RELATIONS BOOKS, ENFORCING RESTRAINT 274 (Lori Fisler Damrosch ed., 1993). For a piece devoted specifically to the problem of collective punishment and the Iraqi sanctions regime, see Michael Sklaire, Note, *The Security Council*

international law and accepted notions of justice, would have it that citizen passivity in the face of violations of international law or even widespread acquiescence in following the commands of the state calls for some type of collective responsibility, the bifurcated compliance and enforcement systems taken together are internally inconsistent. The compliance system, as I have discussed, makes no attempt to pierce the veil of the state to extend its legal obligations to common citizens. Thus, citizens under the present system are in effect held collectively responsible for violating laws that they were not expected to comply with in the first place.

Nowhere has the dysfunction of the international enforcement system become more clear than in the U.N. Security Council-authorized war and the ten-year-long sanctions regime against Iraq. Perhaps indicative of the overall state of the international enforcement system, this is frequently thought of as a great example of successful international enforcement because Iraq was dislodged from Kuwait.[40] During the war, however, hundreds if not thou-

---

Blockade of Iraq: Conflicting Obligations Under the United Nations Charter and the Fourth Geneva Convention, 6 AM. U. J. INT'L L. & POLY 609, 624, 635 (1991). Numerous of the international humanitarian law and human rights conventions prohibit collective punishment. Among provisions in humanitarian conventions, see Article 33 of the Fourth Geneva Convention Relative to the Protection of Civilian Persons in Time of War, opened for signature Aug. 12, 1949, 6 U.S.T. 3517, 3538-40, 75 U.N.T.S. 287, 308-10 (prohibiting punishment of a person "for an offense he or she has not personally committed"). *See also* Convention Relative to the Protection of Civilian Persons in Time of War, *opened for signature* Dec. 12, 1977, Protocol I, art. 75(2)(d), 16 I.L.M. 1391, 1423 (prohibiting collective punishment of persons under the control of a party to a conflict); Convention Relative to the Protection of Civilian Persons in Time of War, opened for signature Dec. 12, 1977, Protocol 11, art. 4(2)(B), 16 I.L.M. 1442, 1444 (prohibiting collective punishments of persons not taking part in hostilities). See generally Report of the International Law Commission on the Work of Its Forty-Third Session, [1991] 2(2) Y.B.INT'L L. COMM'N 1, 104-05, U.N.Doc. A/CN.4/SER.A/ 1991/Add. 1(Part 2) (identifying collective punishment as an exceptionally serious war crime). While the term "collective punishment is not used in international human rights instruments, it is clearly prohibited by core provisions of such instruments. Implicit, for example, in provisions of the Universal Declaration of Human Rights such as the right "to equal protection of the law," (Universal Declaration of Human Rights, art. 7, G.A. Res. 2174 U.N. GAOR, 3d Sess., pt. 1, at 71, U.N. Doc. A/810 (1948) [hereinafter UDHR]), the right to be free of arbitrary arrest, detention, or execution (UDHR, art. 9) and the right "to be presumed innocent until proved guilty" (UDHR, art. 11) is that individuals should not be held accountable for wrongdoings that they did not commit. Other major human rights instruments provide similarly. See International Covenant on Civil and Political Rights, supra note 37, especially arts. 6(1), 901, 14(1)(2)(3), and 26; European Convention on the Protection of Human Rights and Fundamental Freedoms, Nov. 4, 1950, 213 U.N.T.S. 221, especially arts. 5 (l)(a)(b)(c, 6(2)(3); American Convention on Human Rights, Nov. 22, 1969, 1144 U.N.T.S. 123, 9 I.L.M. 99, *especially* art. 5(3) ("punishment shall not be extended to any person other than the criminal").

[40] *See, e.g.,* Bruce Russett & James Sutterlin, *The U.N. in a New World Order*, FOREIGN AFF., Spring 1991, at 69, 75.

sands of Iraqi conscripts were killed,[41] almost all of whom, of course, had no role in the decision to invade Kuwait. While the Iraqi dictator is reported to be profiting from smuggling opportunities made possible by the sanctions regime[42] (which has been ineffective at meeting its political objectives),[43] many Iraqi children are reported to have died as a direct result of the sanctions.[44] Obviously, any enforcement system that systematically sanctions the innocent rather than the infractors is both perversely unjust and deficient in promoting law-abiding behavior.

Compounding the dysfunction of this system where enforcement is almost entirely against states is the symmetrical reality that enforcement is almost entirely by states. Under the bifurcated system, states largely stand between the international order and the real persons whose activities are ultimately necessary to carry out enforcement measures. The international order only maintains direct but limited control over a small number of international civil servants composed mostly of technical specialists. For example, when the United Nations Security Council takes action under Chapter VII, the organization itself has no direct means of enforcing economic sanc-

[41] Estimates of the number of Iraqi soldiers killed have varied widely. The Pentagon's Defense Intelligence Agency estimated in 1991 (with a self-proclaimed error factor of 50% or higher) that approximately 100,000 Iraqi military personnel were killed in action. *See* Barton Gellman, *One Year Later: War's Faded Triumph; Oil Flows, but Saddam Endures and New World Order Is Elusive*, WASH. POST, Jan. 16, 1992, at A1. Other subsequent estimates have been much lower. SEE U.S. NEWS & WORLD REPORT, TRIUMPH WITHOUT VICTORY: THE UNREPORTED OF HISTORY THE PERSIAN GULF WAR ix (1992) (estimating that approximately 8,000 Iraqis were killed); John Heidenrich, *The Gulf War: How Many Iraqis Died?* FOREIGN POL., Spring 1993, at 108, 124 (arguing that number killed may have been as low as 1,500).

[42] *See* Julian Borger, *Iraq: the Next Moves: Saddam's Elite Rides High Despite U.N. Sanctions*, GUARDIAN, Mar. 3, 1998, at 12

[43] *See* Alan Dowty, *Sanctioning Iraq: The Limits of the New World Order*, WASH. Q., Summer 1994, at 179 (reporting that the post-gulf-war U.N. sanctions regime have had "considerable impact on Iraq [in economic terms] without producing the desired political effect").

[44] *See* Sarah Zaidi & Mary Smith-Fawzi, 346 LANCET 1485 (1995) (reporting on the 1995 Food and Agricultural Organization study finding that post-Gulf War sanctions were responsible for the deaths of 567,000 Iraqi children); Barbara Crossette, *Unicef Head Says Thousands of Children are Dying in Iraq*, N.Y. TIMES, Oct. 29, 1996, at A8 (reporting that, according the 1995 UNICEF report, 4,500 children under the age of five were dying every month in Iraq from hunger and disease). These figures have been very controversial. The question has been highly politicized and the methodologies in these studies have been questioned, along with their reliance on official Iraqi information sources. *See* George Lopez & David Cortright, *Counting the Dead,* BULL. ATOM. SCI., May/June 1998, at 39, (1998). Whatever the exact number of dead, however, it has been clear to all observers that the Iraqi sanctions regime has inflicted tremendous hardship on the Iraqi civilian population. See *id.*

tions or applying force against a country.[45] Rather, it must rely upon its state members who are alone capable of mobilizing citizens to carry out enforcement measures. In the Gulf War Chapter VII action against Iraq, the United Nations Security Council "authorized" what was effectively the United States, Britain, France, and Saudi Arabia to "use all necessary means" to get Iraq out of Kuwait,[46] and it is primarily American naval ships that have been enforcing the oil sanctions against Iraq.[47] Practical proposals to enhance the independent enforcement abilities of the international order have mostly fallen on the deaf ears of states (particularly powerful states) committed to maintaining the prerogatives that come with the strict bifurcation of the global order.[48]

Under such a system, the imposition of enforcement measures becomes highly politicized. In determining whether enforcement measures will actually be imposed against an infractor, such legitimate considerations as the severity of a violation, the future danger posed by the infractor, or its past incorrigibility are weighed against the geopolitical and other interests of would-be enforcing states. Strategic alliances, business considerations, the ethnic identifications of domestic constituencies, and the general diplomatic and military power of the infractors can individually or collectively conspire to preclude enforcement in any given case. Not only does the politicization of the enforcement system compromise enforcement in individual cases, but to the extent states can predict that no action will be taken against them, the enforcement system loses its ability to deter violations. Perhaps most importantly, such politicization undermines the perception that law rather than politics guides the international enforcement system, greatly frustrating the creation of an international ethic of respect for the rule of law.

---

[45] *See generally* Martti Koskenniemi, *The Place Of Law In Collective Security*, 17 MICH. J. INT'LL. 455 (1996).

[46] *See* S.C. Res. 678,263 mtg., 29 I.L.M.1565 (1990).

[47] Other countries participating under United States command are: Australia, Belgium, Britain, Canada, the Netherlands, and New Zealand. *See* Barbara Crossette, *Illegal Iraqi Oil Shipments Increase, U.S. Says*, N. Y. TIMES, Nov. 19, 1997, at A8. While the post gulf war weapons inspectors, who were responsible for monitoring Iraqi compliance with the requirement that it eliminate its weapons of mass destruction, worked for the United Nations, even they found themselves relying heavily upon national intelligence agencies for information. *See generally* SCOTT RITTER, ENDGAME: SOLVING THE IRAQ PROBLEM - ONCE AND FOR ALL (1999).

[48] Among such proposals has been one to establish a small standing army under the auspices of the United Nations. *See* David Boren, *The World Needs an Army on Call*, N.Y. TIMES, Aug. 26, 1992, at A21; Alan K. Henrikson, *How Can the Vision of a 'New World Order' Be Realized*, FLETCHER F. WORLD AFF.,Winter 1992, at 63, 76; Editorial, *A Foreign Legion for the World*, N.Y. TIMES, Sept. 1, 1992, at A16.

An empowered Global Peoples Assembly could remedy both of these dysfunctions. If, as suggested, a GPA dedicated to the creation of a fully functional international system were to extend the obligation to comply with transnational law directly to individuals, presumably it would correspondingly provide that that law be enforced against them as well. Then, in accordance with principles of fairness and deterrence, those actually responsible for law-breaking would themselves be the target of legal sanctions, and the ability of states to mobilize their citizens in contravention of the law would be further eroded. In addition, with a GPA, enforcement would no longer have to be dependent upon the vacillating willingness of states to undertake enforcement action. Over time, an empowered GPA would have the potential to directly authorize an international infrastructure to enforce international law directly. While visions of a standing army come to mind, this would be by far the most controversial enforcement mechanism that might come to be considered by the GPA. Most direct enforcement would be far more everyday and mundane. Armies of bureaucrats rather than soldiers would be the most likely to be marshaled and would in many ways provide a far more effective and hopeful vision for the future. To the extent that the bifurcated structure of the international order begins to break down and international law comes to be directly enforced against citizens, the state will begin to lose its ability to break international law. With the growth of a more law-abiding world, it is to be hoped that standing armies will become a much less prominent feature of the global architecture.

III. Fulfilling the Promise: The Empowerment of the Global Peoples Assembly

It is, of course, one matter for a citizen-created Global Peoples Assembly to proclaim internationally prescribed regulation to be binding, applicable to citizens, and directly enforceable on citizens by international institutions. It is quite another matter for such proclamations to become institutional reality. To counter the impression that the creation of an Assembly-led seamless global legal system, while perhaps desirable, is purely in the realm of the utopian, I will now explain how such a transformation could over time become a reality.

As Richard Falk and I explain elsewhere,[49] the ideological underpinnings of the global political order were at one time based upon the divine right of

---

[49] *See generally* Falk & Strauss, *On the Creation, supra* note 1. In many ways the present article takes up where the above referenced piece with Professor Falk leaves off. The

kings. The bifurcation of the international order arose because monarchs who personified the state were thought to be the fundamental repository of all worldly political authority. Under this conceptual structure monarchs were empowered to lord over their subjects below and to become the exclusive progenitors of the inter-state order above. As a result of a fundamental paradigm shift that has occurred, beginning over two hundred years ago, however, political authority is no longer thought to come directly from an absolutist ruler.[50] Rather, people today increasingly accept the notion that political authority comes from the citizens who are called upon to validate political leadership through periodic elections.[51] While the state maintains its role as the final political authority, what is different is that this authority is now delegated from its ultimate source, the citizenry. This means that if citizens were to circumvent the state and themselves create an elected transnational authority to which they would directly delegate certain transnational powers, the state could no longer lay ideological claim to be the sole source of all international political authority. For the first time, an international institution would have a basis of authority separate from that which has been derived from the consent of states.

The Global Peoples Assembly's democratic foundation, however, would not immediately translate into the kind of binding powers that would end the dysfunction of the international system. Traditional political and bureaucratic structures reinforcing state power and the established notion that international law is created by states would not automatically transform themselves with the mere realization of the Global Peoples Assembly. Rather, at first the Assembly as a non-governmental institution would likely be regarded, at least within official circles, as only empowered to issue unofficial and non-binding resolutions and declarations. In what Professor Falk and I refer to as the "socio-political and ideological dynamics of empowerment," however, the Assembly would hopefully come to achieve its democratic potential.[52]

What we mean by socio-political empowerment are those social and political dynamics that would be unleashed by the existence of the Assembly and would enhance its ultimate ability to command authority. This process of

central aim of our joint work is to argue that civil society is now capable (hopefully with the help of certain progressive states) of founding the Global Peoples Assembly, and that because this Assembly would have a basis in popular legitimacy, it would, despite its unofficial origins, have the potential to play a major role in global governance.

[50] *See generally* ROBERT DAHL, ON DEMOCRACY (1998); DAVID HELD, DEMOCRACY AND THE GLOBAL ORDER: FROM THE MODERN STATE TO COSMOPOLITAN GOVERNANCE (1995).

[51] This is, of course, not true everywhere, but the movement toward the belief and practice, though imperfect, of electoral democracy in domestic settings is unmistakable. *See supra* note 2 and accompanying text.

[52] *See* Falk & Strauss, *On the Creation, supra* note 1.

socio-political empowerment would begin with the first elections. Born amidst the publicity that would result from world-wide citizen participation in elections and uniquely capable of speaking in the name of the global citizenry, the Assembly would act as a magnet drawing organized interests with globally oriented policy agendas into its field. Civil society organizations, in particular those that, as I have mentioned, have been frustrated in their attempts to participate in the state-centric international system, would likely take advantage of the democratic opportunity afforded by such an Assembly. By lobbying the GPA to endorse their positions formally, these organizations would force states and others with opposing policy goals to either concede the legitimacy afforded by the only popularly elected global body or to themselves competitively engage its processes.

Our experience of parliamentarianism writ large tells us that as transnational citizen groups organize to petition the Assembly directly, they would form coalitions with other like-minded citizen groups for the purpose of more effectively challenging other similarly coalescing groups.[53] As the organized citizenry increasingly came to find the Global Peoples Assembly useful as its forum for resolving political conflict of a global nature and achieving a workable social consensus, the center of political gravity would gradually shift in favor of the GPA. Allowed for the first time to participate in the international law-making process directly, the organized citizenry would tend to become institutionally committed to the Assembly and invested in its legislative product.

The citizenry at large would as well tend to be drawn into the Assembly's field. The need for candidates for the Assembly to appeal to the electorate would naturally incorporate it into the global system. Discussions that once were considered the elite province of foreign relations specialists and in which citizens could at best participate indirectly at the national level would be as a matter of course broadened to incorporate the newly enfranchised citizenry as a whole. Still this political vitality would not in itself mean that the Assembly would gain the power to create binding law applicable and enforceable in the ways called for in this article. This would potentially result, however, from the addition of what Professor Falk and I call the ideological dynamic of empowerment, the acceptance of the Assembly's democratic claim to exercise legal authority by those existing institutions that are themselves imbued with formal law-making powers.[54]

---

[53] For a case study of the functioning of democratic structures which has become a classic in the study of pluralism, see ROBERT DAHL, WHO GOVERNS? DEMOCRACY AND POWER IN AN AMERICAN CITY (1961).

[54] *See* Falk & Strauss, *On the Creation, supra* note 1.

Domestic courts are the most obvious of such institutions. Certainly at least a few pioneering litigants with the institutional backing of the Assembly's supporters, or simply because such arguments would be helpful to their cause, would begin to cite the Assembly's directives as authoritative. On the pages of law journals, academic supporters of the Assembly could be counted on to provide the rationale for this position by applying globally the familiar arguments supporting the legal powers of democratic assemblies. If the Assembly were in fact gaining the institutional allegiance of the citizenry, some of the bolder progressive judges in open societies would be tempted to be among the first to accept this democratic logic.[55] To the extent judicial acceptance were to occur,[56] the ideological coherence underlying the doctrine of dualism would be eroded, and with such erosion the state's political authorities would lose from within their absolute control over the state's ability to violate international law.

Other law-making institutions could as well help solidify the Assembly's *de jure* legislative authority. Acceptance of the Assembly's powers by international courts and tribunals would put pressure on state executive and political decision-makers to recognize the Assembly's authority. Further pressure could come from the Assembly's efforts to lobby governments on its own behalf as well as from a GPA-organized international plebiscite putting approval of the Assembly's powers to the global citizenry.

Some national parliaments or their equivalents might unilaterally accept the Assembly's powers. However, the definitive event would be when states, the primary entities with a rival claim to power, act collectively to formally recognize the Assembly's legislative authority. This could be done by way of a treaty officially defining the Assembly's powers as part of the overall constitutional structure of the United Nations system of global governance. This, of course, would not happen solely as a result of the compelling logic in favor of the Assembly's democratic authority, but rather out of an acceptance of a new political reality that sees citizens increasingly coming together to resolve political conflict and achieve social consensus within the framework of the Global Peoples Assembly. As this citizen heart of pluralist decision-making migrates in the direction of the Global Peoples Assembly, govern-

---

[55] Of course, the greater the judicial acceptance, the more litigants would begin to make such arguments; which would in turn lead to greater judicial acceptance.

[56] There is some empirical evidence indicating the potential for such acceptance. Some courts have given legal force to resolutions of the United Nations General Assembly. While the General Assembly is composed of almost all of the countries, its formal powers under the United Nations Charter are largely precatory. For further elaboration, see Falk & Strauss, *On the Creation, supra* note 1. The Global Peoples Assembly would be more truly democratic than the General Assembly and, of course, more connected to the citizenry. Professor Falk and I, therefore, conclude that its authority has far more potential for acceptance. *See id.*

ments would only be acknowledging a reality over which they had little direct control.

If a Global Peoples Assembly were to be established at the instigation of civil society, it is, of course, impossible to know if it would develop in the ways suggested. Much would depend upon serendipity and the competence and integrity of the representatives who came to be identified with the Assembly. There are historical reasons, however, to be optimistic that the dynamics I have suggested would play themselves out in ways that would empower the Assembly. After all, the venerable English Parliament began as but an advisory body to the crown. Nevertheless, as the democratic ideal took hold its status as the representative of the people secured its eventual supremacy. Moreover, in our own time there is indeed precedent for a transnational assembly along the lines of a GPA. The directly elected European Parliament, though once a largely symbolic representative of the peoples of the European Union, has today attained significant law-making powers.[57] What is more, it is the national governments of the European Union that have acted primarily on their own initiative to create and strengthen that institution.

Perhaps the most significant force that would propel the eventual empowerment of a GPA is the sheer need for such an organization. The global system is dysfunctional, and as globalization proceeds, the need to find transnational solutions to once local problems continues to accelerate.[58] As the recent protests in Seattle have made clear,[59] citizens are increasingly unlikely to continue willingly to allow themselves to be excluded from the global decision-making process. Political and commercial elites attempting to solve global regulatory problems are, therefore, increasingly likely themselves to come to accept the need for some kind of innovation along the lines of the GPA.

If indeed a GPA does come to pass, and if its authority is to evolve in accordance with some rough approximation of what has been described, then over time, it could carry out what has been called for in this article as necessary to overcome the dysfunction of the international system: make interna-

---

[57] For further elaboration, see FOUNDATIONS OF DEMOCRACY IN THE EUROPEAN UNION: FROM THE GENESIS OF PARLIAMENTARY DEMOCRACY TO THE EUROPEAN PARLIAMENT (John Pinder ed., 1999).

[58] *See* JOSEPH A. CAMILLERI & JIM FALK: THE END OF SOVEREIGNTY? THE POLITICS OF A SHRINKING AND FRAGMENTING WORLD (1992) (discussing the limitations of national political solutions to the global problems of today).

[59] For a discussion of the protest at Seattle's WTO meeting, see e.g., Juliette Beck & Kevin Danaher, Editorial, *Is the WTO a Blessing or A Curse?*, S.F. CHRON., Nov. 29, 1999, at A25.

tional laws that are binding, applicable to citizens, and directly enforceable on citizens.

IV. Conclusion

Some will regard the ability of a popularly elected Global Peoples Assembly to seriously remedy the major dysfunctions of the international system as a hopelessly utopian dream. I have done my best to respond. Others, however, might come to the opposite conclusion. They might see the GPA as a potentially dangerous Leviathan paving the way toward world domination. In concluding, I would like to briefly address this concern.

A word of qualification is first in order. It is far beyond the scope of this article to suggest in any detail the structure of a Global Peoples Assembly. While future academic works exploring alternative models would be useful, the ultimate framework for a GPA must, of course, be determined by democratic process, and at any rate, as the analysis in this article makes clear, a GPA born of citizen initiative would very much be a constitutional work-in-progress. The role it would eventually settle into playing in the overall global order would not be known for many years.

Having said all this, to propose a Global Peoples Assembly endowed with certain law-making powers is not the same as to suggest an all-powerful sovereign-like organization. Such a model would likely be unacceptable to just about everyone. The template that would achieve the most support would likely be drawn from some sort of federalist or confederalist model. In accordance with what the Europeans call "subsidiarity" — the principle by which decisions are made as closely as possible to the citizen[60] — the Assembly's powers would be limited in such a way as to protect autonomy over national matters.[61] Indeed, such deference to internal affairs is the vision presently endorsed by Article 2(7) of the United Nations Charter.[62] Likewise, given the present commitment to international human rights, the international

---

[60] For a discussion of subsidiarity, see Denis Edwards, *Fearing Federalism's Failure: Subsidiarity in the European Union,* 44 AM. J. COMP. L. 537 (1996). *See also* George .A. Bermann, *Taking Subsidiarity Seriously: Federalism in the European Community and the United States*, 94 COLUM. L. REV.331 (1994).

[61] Of course, differences of opinion would present themselves as to exactly how the subsidiarity principle should be applied in concrete situations. This has been the case in Europe and even the United States where debates about federalism have recurrently surfaced.

[62] "Nothing contained in the present Charter shall authorize the United Nations to intervene in matters which are essentially within the domestic jurisdiction of any state. .. ." U.N. CHARTER art. 2, para. 7.

community is likely to insist that an empowered Assembly be limited by the Universal Declaration of Human Rights.[63] How these rights would be enforced as well as the GPA's relationship with the United Nations and other international organizations and tribunals would all have to be determined.

Would there be any absolute guarantees that despite the best of intentions such an Assembly would not help pave the way toward a consolidation of power that could be fundamentally threatening to freedom? Probably not. However, it seems clear from political theory, experience at the national level, and indeed common sense that democratic outcomes are most likely to proceed from democratic structures. In addition to any structural checks and balances established, the Assembly's legitimacy would be grounded in popular democracy and its usefulness as a forum for pluralist decision-making rather than loyalty to a sovereign, an authoritarian ideology or nationalism. Such an institution would be difficult to convert to authoritarian ends without losing the normative command over citizen legal obligations upon which, as previously discussed, the institution's authority ultimately depends.

That the present international order is, in contrast, not structured to allow for democratic participation has been the basic premise of this article. What may be less obvious, however, is the extent to which the bifurcated system is configured to allow authoritarian governance to persist at the national level. In the first place, national governments, whether they be democratic or not, continue in their role as the ultimate and exclusive focal points for citizens' legal obligations. Even to the extent that national governments come to embody the democratic spirit, such a bifurcated system of discrete democracies channels citizen legal obligations so that state-defined societies are structurally juxtaposed to each other. Coercion of one society by another ultimately backed by the ever-present threat (and often use) of military violence is an inherent feature of this bifurcated structure. Such architecture is, of course, the antithesis of seamless democratic governance which results when an integrated democratic structure is designed to protect the rights and security of everyone. To make matters worse, in truth even the best of discrete democracies cannot enjoy an insulated fully functional democratic existence. The democratic health of every society will be to some extent compromised by national security concerns as long as the bifurcated system predetermines an ever-present threat from other societies.

On balance, therefore, it seems far more likely that a democratic future will be attained with the implementation of democratic structures at the international level than without. As this article has suggested, the bifurcated system which provides that states be the ultimate focus of citizen legal obli-

---

[63] UDHR, *supra* note 39.

gations is a remnant from a previous era when absolutist national rulers were accepted as divinely mandated with supreme and absolute power. In the modern democratic view it is commonly held that within states political power at all levels of governance must directly involve the citizenry. It is time these basic democratic principles were applied to the international order. As more political decision-making takes place within the confines of the undemocratic international order, to continue to leave this anomaly unchallenged is to lose ground in the quest for democracy and to see the dysfunction of the bifurcated system become increasingly debilitating.

4
On a GPA as an Alternative to U.S. Hegemony

Toward A Global Parliament

by Richard Falk and Andrew Strauss

The Nation, 2003*

Global sentiment overwhelmingly rejects the Bush doctrine and its antidemocratic assertion of an American right to dictate collective security unilaterally. Faced with the prospect of a looming war in Iraq, millions around the world took to the streets in protest, sadly with little discernible effect. Now, in the aftermath of the war, those who are serious about promoting a world order that is democratic, equitable and sustainable must consider why so much popular energy produced such meager results and how such energy can be more effectively harnessed in the future.

First of all, it is important for peace forces to advance beyond protest and rejectionism. The global peace-and-justice movement urgently requires its own alternative vision. But beyond this, we believe that this is one of those times when concrete steps for global reform should be proposed and acted upon. A positive vision of world order and the future of the United Nations should be as bold in moving toward global democracy as the Bush Administration's vision is in advancing its plans for global dominance.

Specifically, we suggest introducing into the global arena an institution that enables citizens to participate directly in the world political process regardless of their geographic location: namely, a citizen-elected Global Parliamentary Assembly (GPA). The struggle against American unilateralism will gain strength to the extent that the peoples of the world find ways to have their voices heard.

At present, there is no consistently effective way to counter the ability of US leaders (or leaders of any other states, for that matter) to mobilize the citizens and resources of their states for purposes at odds with the rules of international law. The world order remains a global system of states rather than laws when it comes to peace and security: Only when citizens are given an institutionalized site of struggle in the international system and citizen

* Reprinted with permission from THE NATION, September 22, 2003.

politics is allowed to operate beyond the confines of sovereign states is it likely that new sources of authority will gradually emerge.

A GPA would strengthen the international system by creating a new democratic core to that system. Vertically, the global parliament would derive legitimacy and power from its direct, unmediated link to the world's citizenry. And horizontally this new democratic body would be uniquely qualified to oversee and link the currently disjointed system of weak and disparate international organizations. It is important to realize that the UN as currently constituted is a club of states as represented by governments. How different from the Security Council debate on the prospective war against Iraq would have been a discussion representing the strongly held views of citizens.

What we are suggesting is neither a pipe dream nor a grandiose scheme for world government. Its prototype already exists regionally in the form of the European Parliament. Established in 1957, the European Parliament is, along with the Council of the European Union and the European Commission, one of the three lawmaking bodies of the European Union. In the early days, delegates to the Parliament were appointed by national parliaments, but in 1979 citizens began directly electing representatives. Though it started life as a largely advisory body, its character as the direct representative of the European citizenry has created an inexorable momentum toward empowerment. As a result of the Maastricht, Amsterdam Nice treaties over the past decade, the Parliament now has veto power over approximately 80 percent of European Union legislation. Additional powers are envisioned in the constitution for the European Union that is currently under consideration.

Like the European Parliament, a global parliament could start modestly and develop incrementally. It could be established initially in various ways, including as an initiative by a vanguard of democratic governments willing to act as "world order pioneers." As few as twenty to thirty geographically, culturally and economically diverse countries would be enough to credibly launch this experiment in global democracy. So as to defuse resistance from apostles of the status quo, following the example of the European Parliament, its powers could be advisory·during its early years.

Once in place, a global parliament would, it is hoped, over time increase in influence and reputation. The election process would by itself insure a distinctive institutional identity. Citizen groups would be encouraged to petition the global parliament to pass resolutions supportive of their positions. Those opposed to the policy preferences of these citizen groups, whether industrial lobbies, labor unions, states or other citizen groups, would likely be unwilling to concede to their opponents the legitimacy of the only popularly elected global body. Instead, they would likely come to participate as well. It is even possible that nationalistic critics and policy-makers hostile to global democracy would be inclined to participate and put forth their own views. As groups found in the parliament a transnational civic space in which to work out their differences, the center of political gravity could subtly shift in the parliament's favor. Allowed for the first time to participate in the international lawmaking process directly, the organized citizenry would tend to become institutionally committed to the GPA and invested in its activities.

As soon as the parliament begins functioning, citizen groups from countries around the world could exert pressure on their governments to join in the venture. Once a critical mass of membership was reached, even authoritarian governments might find it politically awkward to deny their citizens the right to be represented. At some point in its evolution, the parliament's formal legal powers, as well as its relationship with the UN, would have to be worked out and augmented by a constitutional process. Perhaps it could, alongside the General Assembly, become part of a bicameral global legislative system that would supplement the Security Council as the organ of the UN entrusted with primary responsibility in the area of peace and security. Whatever its precise legal evolution, the process of discovering and legalizing the role of the GPA would itself encourage a worldwide debate on the shape and substance of global governance.

This evolutionary process would take many years, possibly several decades. During this period, the parliament could still exert a benign moral influence that would complement the work of existing civil-society monitors and activists. By holding regional hearings, issuing reports, responding to citizen petitions and passing resolutions, the GPA could gradually introduce a greater measure of popular accountability into existing global institutions and help inform world public opinion about threats to human wellbeing neglected by states.

The mere establishment of a global parliament would be a welcome step, giving hope in a dark time. Taking such a step would signal the emergence of a democratic and peace-oriented alternative to achieving national security through domination and recurrent warfare. In a global parliament, delegates would not represent states, as they do at the United Nations, but rather the citizenry directly. As occurs in other multinational parliaments such as in India, Belgium and the European Parliament itself instead of voting along strictly national and ethnic lines, many delegates would come to vote along lines of interest and ideology. Thus, shifting transnational coalitions seeking the peaceful resolution of international disputes might be able to discourage political leaders and their publics from a reliance on armed conflict, and over time this might slowly lead to the withering away of war as a social institution. At the very least, the global climate would become more receptive to serious disarmament initiatives.

Likewise, the GPA would offer disaffected citizens a constructive alternative to terrorism and other forms of political violence. Those alienated by perceived injustices or by global silence about their grievances would no longer have to choose between surrender and the adoption of desperate tactics. Instead, they would have a legitimate international forum in which they could at least be heard and perhaps find enough support to achieve peaceful redress. Citizens would be able to stand for office, champion candidates and form coalitions to lobby the parliament, a process that would bring those with diverse or opposing views into a give-and-take setting that would improve the chances for compromise and reconciliation. Those whose views did not prevail would likely be more inclined to accept defeat out of a belief in the fairness of the process, and with an understanding that they could continue to press their cause on future occasions.

Of course, the brand of religious extremism associated with September 11 is decidedly antidemocratic in outlook. It is reasonable to question the ability of a parliament to successfully absorb supporters of Al Qaeda and groups with comparable agendas of violence. A salient feature of the liberal parliamentary process at its best, however, has been a capacity to assimilate even those who do not share a pre-existing commitment to democracy. Because a parliamentary process allows for participation and has the ability to confer popular legitimacy on a policy position, experience suggests that even those with extreme agendas will often be drawn into the process-though they may voice dissatisfaction with it and be frequently discouraged by the results. Most aggrieved people in the world, however, are neither ideologically antidemocratic nor naturally prone to extremism and, therefore, given democratic options are much less likely to resort to violence. It is notable that Israeli Arabs continue to value their participation in the Knesset and that Sinn Fein has felt the same with regard to the Northern Ireland Assembly. Of course, the Osama bin Ladens of the planet will never accept the legitimacy of a global parliamentary process. But their ability to attract a significant following might well be substantially diminished by the presence of such an institution, especially if the legitimate grievances of peoples around the world were being consistently addressed with an eye toward the realization of global justice and the promotion of the rule of law.

5
On Terrorism and the Need for a GPA

The Deeper Challenges of Global Terrorism, A Democratizing Response

by Richard Falk and Andrew Strauss[1]

in Debating Cosmopolitics (Archibugi ed.), 2003[*]

Answering the terrorist challenge

The audacious and gruesome terrorist attacks on the World Trade Center and the Pentagon, along with the military response, have been the defining political events of this new millennium. The most profound challenge directed at the international community, and to all of us, is to choose between two alternative visions. What we call the traditional statist response emphasizes 'national security' as the cornerstone of human security. Centralization of domestic authority, secrecy, militarism, nationalism, and an emphasis on unconditional citizen loyalty, to her or his state as the primary organizing feature of international politics are all attributes of this approach.

We recommend an alternative vision, one that we call democratic transnationalism. Democratic transnationalism attempts to draw on the successes of democratic, particularly multinational democratic, domestic orders as a model for achieving human security in the international sphere. This approach calls for the resolution of political conflict through an open transnational citizen/societal (rather than state or market) centred political process legitimized by fairness, adherence to human rights, the rule of law, and representative community participation. The promotion of security for individuals and groups through international human rights law in general, as reinforced by the incipient international criminal court with its stress on an ethos of individual legal responsibility, assessed within a reliable constitutional setting, is a crucial element of this democratic transnationalist vision, which aspires to achieve a cosmopolitan reach.

---

[*] Reprinted from DEBATING COSMOPOLITICS (Daniele Archibugi, ed.) (Verso, 2003).

[1] The authors would like to thank Erin Daly and Daniele Archibugi for their very helpful comments on early drafts of this chapter.

Before the events of September 11 we had argued in favour of the establishment of a distinct, global institutional voice for the peoples of the world as a beneficial next step to be taken to carry forward the transnational democratic project. We proposed a GPA, which we have variously identified as a Global Parliamentary Assembly, and interchangeably as a Global Peoples' Assembly.[2] So far we have deliberately refrained from setting forth a detailed blueprint of our proposal, partly to encourage a wide debate about the general idea, partly to generate a sense of democratic participation in the process of establishing such a populist institution. We have expressed a tentative preference for representation on a basis that would to the extent possible incorporate the principle of one person one vote. The eventual goal would be to enfranchise as voting constituents all citizens of the planet above a certain age. We have further taken the position that the GPA should not interfere in matters appropriately defined as within 'the internal affairs of states', although acknowledging that the extent of such deference is bound to shift through time and often be controversial in concrete instances. The main mission of the GPA would be to play a role in democratizing the formulation and implementation of global policy. It is our conviction that such an assembly's powers should always be exercised in conformity with the Universal Declaration of Human Rights, and other widely accepted international human rights instruments.

We believe that carrying out the transnational democratic project, including establishing the GPA, should be treated as part of the political response to the challenges posed by the sort of mega-terrorism associated with the September 11 attacks. Transnational terrorism, which consists of networks of dedicated extremists organized across many borders, of which al Qaeda is exemplary, is so constituted that its grievances, goals, recruitment tactics and membership, as well as its objects of attack, are all wholly transnational. This form of political violence is a new phenomenon. It is the frightfully dark side of an otherwise mostly promising trend toward the transnationalization of politics. This trend, a result of economic and cultural globalization, has manifested itself in a pronounced way since the street demonstrations staged against the 1999 WTO meeting in Seattle.

The state-centric structures of the international system are not adequate to address this new transnational societal activism and, in fact, the arbitrary

[2] See Richard Falk and Andrew Strauss, 'Toward Global Parliament', *Foreign Affairs*, January-February 2001, p. 212; Richard Falk and Andrew Strauss, 'On the Creation of a Global Peoples' Assembly: Legitimacy and the Power of Popular Sovereignty', *Stanford Journal of International Law*, vol. 36, 2000; Andrew Strauss, 'Overcoming the Dysfunction of the Bifurcated Global System: The Promise of a Peoples' Assembly', *Transnational Journal of Law and Contemporary Problems*, vol. 9, 1999.

territorial constraints on the organization of work and life have intensified various forms of frustration, which feed the rise of transnational terrorism. One cause of this frustration is that globalization in all its dimensions is bringing with it changes of great magnitude that often directly impact on the lives of individuals and regions. These changes range from growing income inequality within and between many societies to powerful assaults on cultural traditions that offend non-Western peoples. Adverse impacts of globalization on many adherents of Islam have definitely induced political extremism in recent decades even before September 11, starting with the Iranian Revolution of the late 1970s.

Even in democratic societies there is a growing sense that domestic politics is not capable of responding creatively to long-range challenges of regional and global scope. It is certainly the case that the magnitude of these challenges is well beyond the capacities of even the strongest of states to shape benevolently on their own. At the same time individuals have an ever-greater incentive to influence global decision-making through their use of the technologies of globalization, especially the Internet. Information technology has given individuals an unprecedented ability to increase their leverage on public issues by making common cause with like-minded others without regard to considerations of geography or nationality.

An institutional framework such as that which would be provided by a GPA is a democratic way to begin peacefully to accommodate this new internationalization of civic politics. Individuals and groups could channel their frustrations into efforts to attempt to participate in and influence parliamentary decision-making as they have become accustomed to doing in the more democratic societies of the world. Presently, with trivial exceptions, individuals, groups and their associations are denied an official role in global political institutions where decision-making is dominated by elites who have been officially designated by states. Intergovernmental organizations, such as the United Nations, the World Trade Organization and the International Monetary Fund are run as exclusive membership organizations, operated by and for states. With the possibility of direct and formalized participation in the international system foreclosed, frustrated individuals and groups (especially when their own governments are viewed as illegitimate or hostile) have been turning to various modes of civic resistance, both peaceful and violent. Global terrorism is at the violent end of this spectrum of transnational protest, and its apparent agenda may be mainly driven by religious, ideological and regional goals rather than by resistance directly linked to globalization. But its extremist alienation is partly, at the very least, an indirect result of globalizing impacts that may be transmuted in the political unconscious of those so afflicted into grievances associated with cultural injustices.

In addition to helping provide a non-violent and democratic channel for frustrated individuals and groups to affect meaningfully global decision-making, a GPA has the potential to provide a way of helping to resolve inter-societal and more recently intercivilizational conflict and polarization. Presently, the institutions around which citizen politics is formally structured are confined within distinct domestic political systems. This makes a unified human dialogue on issues of shared concern impossible. And transnational remedies for perceived injustices are not available. In a globalizing world it is crucial to encourage debate and discussion of global issues that builds consensus, acknowledges grievances, and identifies cleavages in a manner that is not dominated by the borders of sovereign territorial states, or even by innovative regional frames of reference as in Europe. As a consequence of this existing pattern of fragmentation in the political order, societies and cultures develop their own distinctive and generally self-serving distortions and myths, or perhaps, at the very least, experience exaggerated differences of perception that feed pre-existing patterns of conflict. Most persons within one society have little difficulty identifying the distorted perceptions of others, but tend to be oblivious to their own biases, an insensitivity nurtured by mainstream media especially in the midst of major crises. The oft heard American response to the September 11 attacks, 'Why do they hate us so?' and the seething anger in the Muslim world that has risen to the surface in the aftermath of the attacks starkly demonstrate just how profound and tragic is the perception gap for societies on both sides of this now crucial civilizational and societal divide.

The establishment of a GPA provides one way to address constructively this perception gap. Like all elected assemblies, a GPA would be a forum engendering debate on the main global controversies, especially as they affect the peoples of the world. Because elected delegates would be responsive to their respective constituencies, and because the media would cover proceedings, this debate would likely exert an influence far beyond the parliamentary chambers. Its echoes would be heard on editorial pages, listservs, and TV, in schools and churches, and in assorted discussions at all levels of social interaction around the world. Spokespersons directly connected to aggrieved groups of citizens would have a new transnational public arena to voice their opinions and grievances, as well as to encounter opposed views. Those attacked or criticized would have ample opportunity to defend themselves and express their counter-claims. From such exchanges would come the same pull toward a less confrontational understanding between diverse groups of citizens that we find within the more successful domestic democratic systems of the world. Of course, complete agreement would never be achieved and is not even a worthy goal. Conformity of outlook is never healthy for a political community, but it is especially inappropriate in a global setting, given the unevenness of economic and cultural circumstances that exist in the world. But a GPA process could at least greatly facilitate convergent perceptions of reality, thereby making controversies about problems and solutions more likely to be productive, including a mutual appreciation and acceptance of differences in values, priorities and situations.

In addition to helping reduce the perception gap as an underlying cause of social tensions, a GPA would further promote the peaceful resolution of enduring social tensions by encouraging reliance on procedures for reaching decisions based on compromise and accommodation. Even where mutually acceptable solutions are not immediately achievable, parliamentary systems of lawmaking and communication, if functioning well, at least provide a civil forum where adversaries can peacefully debate and clarify their differences. If such institutions generate community respect and gain legitimacy, then those who do not get their way on a particular issue will be generally far more inclined to accept defeat out of a belief in the fairness of the process and with an understanding that they can continue to press their case in the future.

Of course, the brand of Muslim fundamentalism embraced by Osama bin Laden is illiberal and anti-democratic in the extreme. Given the existence of such extremism, it is appropriate to question the ability of liberal democratic institutions to absorb successfully those who share the worldview of al Qaeda, or adhere to similar orientations. One of the impressive features of liberal parliamentary process, however, is its considerable ability to assimilate many of those who do reject its democratic outlook. Because parliamentary process invites participation and because it has the politically powerful capacity to confer or deny the imprimatur of popular legitimacy upon a political position, experience at the domestic level suggests that even those with radical political agendas will seldom decline the opportunity to participate. In the United States, for example, those on the Christian right who have deep religious doubts about the validity of secular political institutions have not only participated in the parliamentary process, but have done so at times with zeal, tactical ingenuity, and considerable success despite their minority status. In other countries, small political parties at the margins of public opinion often exert disproportionate influence in situations where a majority position is difficult for dominant parties to achieve. By participating in the process they have come to accept, at least in practice, the legitimacy of these institutions and procedures for societal decision-making.

Somewhat analogously, in the Cold War era the orthodox Soviet-inspired critique of the American system nominally accepted by those American Communists represented by the Communist Workers Party included a rejection of 'bourgeois' rights in favour of what was then identified as 'the dictatorship of the proletariat'. Yet, despite their professed rejection of 'bourgeois democracy', their leader Gus Hall ran for President of the United States repeatedly in an attempt to gain a tiny bit of electoral legitimacy for his position of isolation at the outermost reaches of public opinion. The relative domestic openness of the American political process helps explains why the United States has suffered relatively little indigenous political violence in the twentieth century. During the period of heightened political tensions in the 1960s, groups committed to violence such as the Weather Underground, unlike al Qaeda today, could not attract popular support for their radical rejection of the American governing process, and never became more than a nuisance, posing only the most tangential threat to the security, much less the stability, of the country. This lack of societal resonance soon leads to the decay, demoralization and collapse of such extremist groups, a dynamic of rejection that is far more effective in protecting society than law enforcement is even if enhanced by emergency powers as is the case in wartime conditions. To a lesser extent, the same self-destruct process seems to have kept the right-wing militia movement from posing a major threat to civic order, although it was indirectly responsible for inspiring the 1995 Oklahoma City bombing. This phenomenon with variations can he observed within all of the more democratic systems of the world. The Osama bin Ladens of the planet would be highly unlikely themselves to participate in a global parliamentary process, but their likely ability to attract any significant following would be substantially undermined to the extent that such an institution existed and gave the most disadvantaged and aggrieved peoples in the world a sense that their concerns were being meaningfully addressed. Indeed, if such a safety valve existed, it might prevent, or at least discourage, the emergence of the Osama syndrome, that the only way to challenge the existing arrangement of power and influence is by engaging in totalizing violence against its civilian infrastructure.

Civic activism: setting the stage for a GPA

We believe that the underlying preconditions for a GPA are being created by the way that civic politics is increasingly challenging the autonomy of the state-centric international system. In one of the most significant, if still under-recognized, developments of the last several years, both civic voluntary organizations and business and financial elites are engaged in creating parallel structures that complement and erode the traditionally exclusive role of states as the only legitimate actors in the global political system. Individuals and groups, and their numerous transnational associations, rising up from and challenging the confines of territorial states, are promoting 'globalization-from-below', and have begun to coalesce into what is now recognized as being a rudimentary 'global civil society'. Business and financial elites, on their side, acting largely to facilitate economic globalization, have launched a variety of mechanisms to promote their own preferred global policy initiatives, a process that can be described as 'globalization-from-above'. While these new developments are rendering the territorial sovereignty paradigm partially anachronistic, they are still very far from supplanting the old order, or even providing a design for a coherent democratic system of representation that operates on a truly global scale. Until the international community creates such a representative structure, the ongoing tension between the democratic ideal and the global reality will remain unresolved. And we will continue to be plagued by an incoherent global political structure in which the peoples of the world are not offered the sort of democratic alternative to violence that is increasingly considered the *sine qua non* of legitimate domestic governance.

The organizations of global civil society

Is this coalescence of personal initiatives with an array of transnational initiatives that we identify as global civil society capable of . mounting a transformative challenge to the customary role of states as the representatives of their citizens in the international system? Civil society, roughly defined as the politically organized citizenry, is mostly decentralized, broken down into non-profit organizations and voluntary associations dedicated to a wide variety of mostly liberal, humanitarian and social causes (though some decidedly illiberal and anti-liberal, of which terrorist and criminal networks are the worrisome instance). Transnationally, the largest and most prominent of these organizations bear such respected names as Amnesty International, Greenpeace, Oxfam, and the International Committee of the Red Cross. There are now more than 3,000 civil society organizations either granted consultative status by the United Nations Economic and Social Council or associated with the UN Department of Public Information.

As described by Jessica Mathews in her landmark 1997 article in *Foreign Affairs*,[3] global civil society gained significantly in influence during the second half, and particularly the last quarter, of the twentieth century. The early 1990s, however, was the time when civic transnationalism really came of age. If any single occasion deserves to be identified with the emergence of civil society on the global scene it would probably be the June 1992 UN Conference on the Environment and Development held in Rio de Janeiro.

---

[3] Jessica T. Mathews, 'Powershifts', *Foreign Affairs*, January-February 1997. p.50.

More than 1,500 civil society organizations were accredited to participate and 25,000 individuals from around the world took part in parallel NGO forums and activities. Civic associations and their representatives were for the first time recognized as an important and independent presence at a major world inter-governmental conference. The Rio Conference, partly responsive to this active involvement of global civil society, produced four major policy-making instruments.[4]

After Rio the pattern intensified. In the first half of the 1990s there were several other major global conferences under UN auspices at which civil society participation was an important factor. The most significant of these dealt with human rights (Vienna 1993), population (Cairo 1994), and women (Beijing 1995). The democratizing success of these global events produced a backlash among several major governments, especially the United States. The result in the short term has been the virtual abandonment of such conferences by the United Nations, supposedly for fiscal reasons, but actually because governments were afraid of losing some of their control over global policy-making. With the exception of the racism conference in Durban, South Africa, during 2001, there has been no major conference of this sort in the new millennium. It is important to evaluate this experience in the setting of the quest for global democracy. There is little doubt that these conferences in the 1990s did a great deal to establish the role and presence of civil society as a significant player in the global arena.

When the 1990s came to an end, the decade's balance sheet of accomplishments reflected for the first time in history the impact of global civil society. These transnational forces had been instrumental in promoting treaties to deal with global warming, establish an international criminal court and outlaw anti-personnel landmines. These same actors were also influential during these years in persuading the International Court of Justice to render an Advisory Opinion on the legality of nuclear weapons and in defeating an OECD attempt to gain acceptance for a multilateral investment agreement. This global populist movement at the turn of the millennium gained widespread attention through its advocacy of the cancellation of the foreign debts of the world's poorest countries. While all of these efforts to a greater or lesser extent remain works-in-progress, civil society has clearly been indispensable in achieving current levels of success.

During the formative years of the 1990s the most visible gatherings of civil society organizations took place beneath the shadow of large multilateral conferences of states. As the decade drew to a close, and with these con-

---

[4] These were: on sustainable development, the Rio Declaration and Agenda 21; to help safeguard the planet's biodiversity, the Biodiversity Convention; and perhaps most importantly, to combat the warming of the planet, the Climate Change Convention.

ferences, at least in the near term, mostly foreclosed, something different began to occur. The multitude of global civil society organizations began to act on their own, admittedly in an exploratory and highly uncertain fashion, and yet independently of states and international institutions. For instance, in May 1999 at The Hague Appeal for Peace, 8,000 individuals, mostly representing civil society organizations from around the world, and given heart by the presence of such luminaries as Nobel Peace Laureates Archbishop Desmond Tutu, Jose Ramos-Horta, and Jody Williams, met to shape a strategy for the future and to agree on a common agenda. Throughout the following year there were similar though smaller meetings in Seoul, Montreal, Germany, and elsewhere.

These meetings were a prelude to the Millennium NGO Forum held at the UN in May 2000 at the initiative of UN Secretary-General Kofi Annan. It was an expression of his 'partnership policy' to reach out to non-state actors of both a civic and a market character. The Secretary-General invited some 1,400 individuals from international civil society to UN Headquarters in New York to present their views on global issues and to debate an organizational structure that might enable the peoples of the world to participate effectively in global decision-making. That UN Millennium Forum agreed to establish a permanently constituted assembly of civil society organizations called the Civil Society Forum that is mandated to meet at least every two to three years, scheduled so as to precede the annual sessions of the UN General Assembly. While progress has been uneven, civil society has been continuing to work in the face of statist resistance and skepticism to bring this forum into fruition.

Many activists within global civil society regard this UN millennial initiative as the first step toward the establishment of a popular assembly that would meet at regular intervals, if not on a continuous basis. The emergence of such a Civil Society Forum might over time come to be recognized as an important barometer of world public opinion, and significantly, from the perspective of this chapter, could be seen as a preliminary, yet significant, step on the path to the establishment of a GPA.

The global business elite at Davos

Complicating, yet undeniably crucial to the dynamics of global democratization, are the efforts of business and finance to reshape the international order to render the global marketplace more amenable to the expansion of trade and investment. Transnational business and financial elites have so far clearly been more successful than civil society. Through their informal networks and their stature in society, financial and business elites often blend seamlessly with national and international structures of governance. State emissaries to the international system are frequently chosen directly from their ranks, and the acceptance of the neo-liberal economic ideology as tantamount to the official ideology not only of international economic institutions, but of other international organizations and most governments, has given business and banking) leaders an extraordinary influence on global policy. Even in formerly exclusive arenas of state action, these private sector actors are flexing their muscles. As an indication of this expanding international influence, by bringing business and banking officials into United Nations policy-making circles for the first time, the UN Secretary-General has made 'partnering' with the business community a major hallmark of his leadership. The United Nations has now established a formal business advisory council that is meant to institutionalize a permanent relationship between the business community and the UN, as well as initiated a 'Global Compact' in which major multinationals sign on to a set of guidelines that commits them to uphold international standards pertaining to environment, human rights and labour practices in exchange for being given what amounts to a UN stamp of approval for their conduct.

As with civic groups, elite business participation in this emerging globalism is in the process of transforming itself into an informal institutional structure that indirectly challenges the statist paradigm. The best example of the ability of elite business networks to extend their influence into the international system has been the' World Economic Forum that has been meeting annually in Davos, Switzerland. The WEF was begun modestly three decades ago by the Swiss business visionary, Klaus Schwab. During its early years the WEF concentrated its efforts primarily on rather humdrum management issues. In the early 1980s, however, it succeeded in transforming itself into a political forum. In many ways Davos as we know it today is the legacy of earlier attempts to create transnational networks tasked with joining together international corporate and policy-making elites. Most observers agree that the most prominent of these precursors to the WEF was the highly secretive Bilderberg Conferences. Also important was David Rockefeller's Trilateral Commission (which also began in the 1970s, with an immediate display of influence on the highest levels of governmental decision-making in the industrial countries of the North before largely fading out of sight, in large part because Western governments adopted and acted upon its policy agenda). In terms of sheer concentration of super-elites from around the world, however, there has never been anything approaching the scale and salience that has been achieved by Davos over the course of the late 1990s. Annually, 1,000 of the world's most powerful executives and another 1,000 of the world's senior policy-makers participate in a week of roundtables, discussions, lectures and presentations by world leaders.

But Davos has become much more than an assemblage of the rich and famous, although it is far less menacing and conspiratorial than its most severe critics allege, and it espouses no grandiose project that seeks to rule the world. At the same time, its advocates often make claims that stretch the reality of its considerable influence beyond the point of credibility. The WEF provides flexible arenas for discussion and recommendation that give its membership the ability to shape global policy on a continuous and effective basis. It is notable that the UN Secretary-General's ideas about a partnership with business and civil society have been put forward as proposals during several high-profile appearances by Kofi Annan at Davos. In addition to encouraging the development of its own well-articulated approaches to global problems on the basis of neo-liberal precepts, the WEF conducts and disseminates its own research, which not surprisingly exhibits a consistent economistic outlook that portrays the future as market-driven. The WEF produces an annual index ranking the relative economic competitiveness of all countries in the world, which is given substantial media attention at the time of release each year.

There is no objective way to gauge the extent of influence exerted by Davos. Its own claims as a facilitator of conflict resolution are often not convincing. For example, the WEF takes credit for facilitating early meetings between the apartheid regime and the ANC, and for bringing Israeli Prime Minister Shimon Peres and PLO Chairman Yasser Arafat together in 1992, where they purportedly reached a preliminary agreement on Palestinian administration of Gaza and Jericho. The WEF is far more discreet about claiming any direct influence on global social and economic policy, being sensitive to accusations of back channel lobbying on behalf of transnational corporate interests. If the focus is placed on global economic policy then Davos together with other overlapping networks of corporate elites, such as the International Chamber of Commerce, seems to have been remarkably successful up to this time in shaping the global policy setting in directions to its liking. This success is illustrated by the expansion of international trade regimes, trends toward privatization, the maintenance of modest regulation of capital markets, the credibility accorded only to a neo-liberal interpretation of state/market relations, and the supportive collaboration of most governments, especially those in the North.

All in all the WEF has managed to position itself so as to provide a vital arena of inquiry and decision during this early stage of economic globalization. Such positioning has reduced the significance of democratic forces operating within states in relation to foreign economic policy, which in turn strengthens the argument to provide opportunities for civic participation in transnational institutional settings that will offset the impact of the multinational corporate arenas and give more voice to grassroots .and populist concerns. Again, the focus on this dynamic is likely to be lost in the short-term aftermath of the September 11 attacks, which has temporarily restored the state as guardian of security to its traditional pre-eminence. Underlying globalizing trends are likely with the passage of time, however, to reassert the significance of establishing the structures of global governance in forms that take into account the goals of both market and transnational civic forces.

A GPA as the logical outcome of the process of global democratization

Putting aside the backlash against the global conference format, it seems reasonable to suggest that the international system is now exhibiting greater participation by non-state actors than ever before in its history. Without question, global civil society is unable to equal the influence, resources and power linkages of the corporate and banking communities. Nevertheless, relying on imagination and information, many of these civic networks have found ways to carve out a niche within the international order that enables effective pressures to be mounted. At the same time, there are many short-comings of such an ad hoc and improvised approach to global democracy. This transformation of the international system has been occurring in a largely uncoordinated and uneven fashion that further tends to disadvantage the concerns of the weakest and poorest. This obscures the need to connect these two types of globalizing networks (from above, from below) in a manner that is coherent, fair and efficient from the perspective of global governance.

In effect, what we have at present is a partial transplant from domestic political systems where interest group pluralism flourishes within an overarching representative structure of parliamentary decision-making. At the global level we currently have rudimentary interest group pluralism, but it is deficient in several respects. There is a lack of accountability due to the absence of a representative structure and a low quality of functionality as a result of statist unwillingness to provide institutional capabilities for transnational political life. We believe this to be an inherently unsustainable path to a more evolved global system that is humane and comes to approximate a democratic model. What is notably missing from these intersecting forms of transnationalism is some type of unifying parliamentary body that can represent general as well as special interests.

The prevailing understanding of democracy today rejects the view that organized interest groups can validly claim to represent society as a whole. As global civil society has become more of an international presence, those opposing its agenda and activism have already begun to ask upon what basis are those within it entitled to represent the peoples of the world. Awkward questions are asked: 'Who other than themselves do civil society organizations speak for?' 'Who elected them?' 'To whom are they accountable for their actions?' As global civil society becomes more influential, and as more ideologically diverse and antagonistic groups such as, for example, the American National Rifle Association, or for that matter Islamic fundamentalist organizations, clamour for access to global arenas of decision, this problem of representation can only become more complex and ever more hotly contested.

This illegitimacy charge can be equally levelled at the Davos improvisation, which, unlike civil society, does not even possess that degree of representativeness that comes from having within its ranks large membership organizations. Certainly those citizens who oppose mainstream globalization regard the Davos model of elite politics to be extremely suspect. Such an assessment of these transnational developments suggests that the kinds of opening of the international system that have been occurring do not satisfy the demand for democratic participation. Something more is needed. Some sort of popular assembly capable of more systematically representing the diverse peoples of the world is necessary if the democratic deficit is to be meaningfully reduced. To the extent that the global undertakings are criticized for their failure to measure up to modern democratic standards, then world order seems ever more vulnerable to the charge of being more of an insiders' game than all but the most corrupt and draconian domestic political systems. Even before the events of September 11 it was evident that those whose interests were not being addressed, were unwilling to accept the legitimacy of existing global arrangements. It seems likely that given the continuation of these conditions, that the democratic deficit will grow even larger, leading to the further proliferation of various types of severe instability, which are currently causing such widespread turmoil and suffering in the world system.

The absence of a unifying parliamentary structure also means that there is currently no institutional vessel capable of bringing together the organized groupings of transnational activism that are identified with civil society and the Davos constituencies so as to facilitate dialogue, and the search for compromises and accommodations. As matters now stand, only governments have the institutional capacity to find such common ground and strike deals. As we discussed previously, there is no process for individuals and groups themselves to create a social consensus across borders or to engage formally with those acting on behalf of market forces. To the extent that solutions to global problems can be arrived at within a structure that institutionalizes interaction and allows for direct communication among competing interests, such interests will be much more likely to accept as legitimate, policy outcomes that have been fairly negotiated and agreed upon.

A GPA as a practical political project

We believe that the establishment of some sort of parliamentary assembly is necessary to begin to deal seriously with the democratic deficit. At the same time we realize that scepticism is rampant: is the creation of such a global assembly politically possible at this stage of history? For a variety of reasons, we believe that it is not Panglossian to believe it possible for the global community to take this vital step in building global democracy. After all, empirically suggesting the viability of such a project is the European Union, which has been making impressive attempts to overcome a purported regional democratic deficit. The EU already possesses a transnationally elected legislative body, the European Parliament. The European Parliament, along with the European Council and the European Commission, is one of three legislative bodies operating within the framework of the European Union. As we would expect to be the case with a globally elected assembly, the Parliament has struggled to establish credibility over time in the face of statist scepticism and media scorn. In recent years, however, the European Parliament has finally begun to gain respect, and has started to exercise significant power. Europe is, of course, far more homogeneous and economically integrated than the world at large, and the establishment of the Parliament was a part of a broader movement toward regional unity. At the same time this European evolution shows that there are no absolute political or logistical barriers to the creation and functioning of such an assembly on a transnational scale, and further, that such a development is fully compatible with the persistence of strong states and robust nationalist sentiments. In fact, on a global level, those with a pronounced interest in global governance — civil society, the corporate elite, and many governments — have an individual as well as collective stake in erecting some type of overarching democratic structure.

The role of civil society

Certain sectors of civil society in particular could likely be, and in fact are being, mobilized to lead the drive for such an assembly.[5] This is important,

---

[5] While still in their early stages, we believe various initiatives merit notice. An organization called the Assembly of the United Nations of Peoples has attempted to bring civil society organizations together into a quasi-representative assembly. In the fall of 2001 it included civil society organizations from the majority of the world's countries in its fourth assembly in Perugia. Italy. Also notable is the Global Peoples' Assembly Movement. This organization had its first major assembly in Samoa in April 2000. Like the Perugia initiative its purpose is to model a globally democratic institutional structure that would enable the

because while there is the potential to find some support from corporate and political elites, it is unrealistic to expect the main initiative to come from these sectors. Most of the individuals leading business and governmental organizations tend to be institutionally conservative, as well as often too closely linked to state structures to support such a bold initiative. For these reasons, the primary energy for a global parliament will come from civil society, or nowhere.

It is rather obvious, however, that not even all civil society organizations are in favour of the creation of such an assembly. Some evidently sense that their influence would shrink in an altered world order. Nevertheless, the sentiments throughout global civil society are overwhelmingly in favour of establishing institutions and practices that will enable global democracy to flourish in the years ahead. Within this broader consensus there exists a realization that the creation of a functioning global parliament or assembly is a necessary and desirable step. The appeal of the GPA proposal to advance the agenda of global civil society seems rather obvious. At a general level, a democratically constituted assembly would be likely to address widespread societal concerns about the undemocratic nature of existing international institutions such as the World Trade Organization, the International Monetary Fund and the World Bank. It would almost certainly encourage further democratizing global reforms, as well as provide a setting for debates about the positive and negative effects of globalization. There would for the first time a widely recognized global forum in which such matters of public be concern as environmental quality, labour standards, and economic justice could be discussed from a variety of perspectives, including encounters between civil society representatives from North and South who set forth contrasting concerns embodying differing priorities. The presence of democratic structures does not, of course, guarantee that participants will consistently behave responsibly. We have learned from experience that even the most experienced and respected legislative institutions within states can act in an erratic fashion from time to time that does not reflect the real interests or values of constituents, but such is the cost incurred to sustain democratic processes as the basis of governance.

Even an initially weak and controversial global assembly could at least provide the beginnings of democratic oversight and accountability for the international system. The fact that individuals from many parts of the world

peoples of the world to have a meaningful voice in global governance. Also worthy of attention are efforts by an organization called Citizen Century to link the national parliamentarians of the world together through the Internet into what it calls a 'Global E-Parliament' and efforts by The World Citizen Foundation to promote the establishment of a globally elected parliament.

would directly participate in elections ' would likely lead the assembly to have an impressive grassroots profile that would lend a certain populist authenticity to its pronouncements and recommendations. In all probability, at first, most governments would refuse to defer to such an assembly that operated beyond their control, but such rejectionist attitudes would be unlikely to persist very long. After all, we are living at a time when, democracy has increasingly become the sine qua non of legitimacy around the world and the assembly would be the only institution that could validly claim to represent the peoples of global society directly. The comparison of its views with those of governments and market-dominated forums would likely attract media attention before long; becoming a part of public discourse would in turn influence the course of civil-political decision-making.

Besides exercising a democratic influence on the formulation of social policy, such an assembly could also be instrumental in helping to encourage compliance with international norms and standards, especially in the realm of widely supported human rights. Currently, the international system generally lacks reliable mechanisms to implement many of its laws. Civil society organizations such as Amnesty International, and even international organizations such as the International Labor Organization and the UN Human Rights Commission, attempt to address this deficiency and exert significant pressure on states by exposing failures of compliance by states, relying on a process that is often referred to as the 'mobilization of shame'. This pressure is premised on the importance to governments of sustaining their reputation for acting in conformity with normative standards and the reliability of established NGOs in identifying patterns of abusive behaviour. In contributing to such an oversight function, a popularly elected GPA would likely soon become more visible and credible than are existing informal watchdogs that seek to expose corporate and governmental wrongdoing, and in any event, would complement such activism. A GPA would also tend to be less deferential to leading sovereign states than the more official watchdogs that function within the essentially statist framework of the United Nations System.

Perhaps most fundamentally, the mere existence and availability of the assembly would likely be helpful in promoting the peaceful resolution of international conflicts. We have already discussed how a GPA might be useful in undermining wider circles of societal support for international terrorism as a form of non-state violence. It could also in time help to reduce the likelihood of interstate violence as well. Instead of representing states, as in the United Nations and other established international organizations, delegates would directly represent various constituencies with societal roots. This means that, unlike the present system, the assembly would not be designed to reinforce artificially constructed 'national interests' or to promote the special projects of rich and influential elites. Rather, as in multinational societies such as India or Switzerland, or in the European Parliament, most elected delegates do not consistently or mechanically vote along national lines, except possibly in instances where their national origins are directly engaged with the issue in dispute. Coalitions form in these settings on other bases, such as worldview, political orientation, class and racial solidarities, and ethical affinities. The experience of engaging in a democratic process to reach legislative compromises on the part of antagonists that are organized as opposing, but non-militarized and often shifting, coalitions may over time help establish a culture of peace. It is perhaps too optimistic to think that such a learning curve might eventually undermine reliance on the present war system to sustain national and global security. It is difficult to transform the militarist mentalities associated with the pursuit of security in a world that continues to be organized around the prerogatives of sovereign national units that are heavily armed and disposed to destroy one another if the need arises. The hope is that over time the organization of international relations would come more closely to resemble decision-making within the most democratic societies of the world. Not only would an assembly tend to oppose military establishments as the foundation of global security, but it is also likely to build confidence in the perspectives of human security and in the efficacy of peaceful approaches to world order. Only when enough people begin someday to feel that non-violent structures of governance, including law enforcement, can ensure their individual and collective survival will meaningful disarmament become a genuine political option.

Any proposed institution that can credibly claim a potential for advancing causes as central to the agenda of various global civil society organizations as global democratization, labour and environmental regulation, effective global governance, peace, and human rights obviously should possess the capacity to generate broad-based support within civil society. So far, however, the nascent civil society movement that favours the establishment of such an assembly remains separate and distinct. It has not managed to gain significant levels of support, or even interest, from the issue-oriented actors that have so far been the main architects of global civil society. The present movement for an assembly consists mainly of individuals and groups who believe in holistic solutions to global problems, and seek to promote humane global governance for the world. Such proponents of a GPA are culturally influenced by a range of contemporary traditions of thought and modalities of action as varied as ecology, religion, spirituality, humanism and, most recently of all, the Internet. Each of these orientations proceeds from a premise of human solidarity and a belief in the essential unity of planet earth. Significant organizing efforts associated either with building support for the GPA or experimenting with its local enactment are under way in many different places around the world. This is an exciting development. It portends the possibility that from within civil society a truly innovative and visionary politics is beginning to take shape after centuries of dormancy. Such movement is an expression of the increasing robustness of democratic values as the foundation for all forms of political legitimacy regardless of the scale of the unit of social action being appraised. Also relevant are many types of transnational connectivity that manifest the globalizing ethos of our twenty-first-century world.

The receptivity of the business elite to a GPA

The global outlook of the corporate and financial elites represented at Davos, and elsewhere is also relevant to the prospects for furthering the cause of a GPA. The Davos network has been singularly successful in marshalling support for new international regimes that promote its interests in an open global economy. The World Trade Organization and NAFTA are two obvious examples. Certainly some within its ranks will oppose a new global parliamentary institution because a more open political system would mean a broader decision-making base, a questioning of the distribution of the benefits and burdens of economic growth, and more pressure for transnational regulation of market forces. Such developments would almost certainly be viewed with suspicion, if not hostility, by those who meet regularly at Davos to construct a world economy that is committed to the 'efficient' use of capital, and dubious about any incorporation of social and normative goals into the formation of world economic policy. It would almost certainly be the case that such an assembly, if reflective of grassroots opinion around the world, would be highly critical of current modes of globalization, and hence at odds with the outcomes sought by the Davos leadership. But with transnational corporations having been, and in all likelihood continuing to be, beneficiaries of this globalization-from-above, those in the business world with a more enlightened sense of their long-term interest are already coming to believe that the democratic deficit must be addressed by way of stakeholder accommodations. It is perhaps relevant to recall that although hostile at first, many members of the American managerial class came under the pressure of the Great Depression and its societal unrest to realize that the New Deal was a necessary dynamic of adjustment to the claims of workers and the poor during a crisis time for capitalism. The same kind of dynamic made social democracy acceptable to the business/financial leadership of leading European countries, and helped give capitalism a more human face that enhanced its legitimacy at the level of society. In a similar vein, many of the leading figures in world business seem to find congenial the idea that some sort of democratizing improvisation along the lines we are suggesting is necessary to make globalization politically acceptable to more of the peoples of the world.

As the large street protests of the last few years in various places around the world suggested to many observers, globalization has not yet managed to achieve grassroots acceptance and societal legitimacy. Lori Wallach (the prime organizer of the Seattle anti-WTO demonstrations) said in an interview that her coalition of so many diverse groups, in addition to battling a series of distinct social issues, was held together by the 'notion that the democracy deficit in the global economy is neither necessary nor acceptable'.[6]

In fact, the main basis of popular support for globalization at present is not political, but economic. Globalization has either been able to deliver or to hold out the promise of delivering the economic goods to enough people to keep the anti-globalization forces from gaining sufficient ground to mount an effective challenge against it. Economic legitimacy alone is rarely able to stabilize a political system for long. Market-based economic systems have historically undergone ups and downs, particularly when they are in formation. The emerging-markets financial crisis that almost triggered a world financial meltdown in 1997 will surely not be the last crisis to emerge from the current modalities of globalization. Future economic failures are certain to generate strong and contradictory political responses. We know that standing in the wings, not only in the United States but in several other countries, are politicians, ultra-nationalists, and an array of opportunists on both the left and the right who, if given an opening, would seek to dismantle the system so as to restore territorial sovereignty, and with it, nationalism and protectionism. If the globalizing elite is seeking to find a political base that will allow it to survive economic downturns, particularly in the event that economic and social forces in powerful countries are in the future adversely affected, then it would do well to turn its attention urgently to reducing the global democratic deficit. Global terror plays a diversionary role at present, especially in the United States, but this distraction from the imperatives of global reform are not likely to persist, especially in the face of widespread economic hardship and distress.

There is a lesson to be learned from Suharto' s Indonesia that offers some striking parallels to the vulnerabilities of the current global system. Indonesian citizens had come to believe in democratic practices, but the political system remained largely authoritarian, and unresponsive to the concerns of the people. As long as Indonesia was both a Cold War ally of the West and enjoyed the dramatic economic growth rates that had been sustained for nearly 30 years, American support was solid and there were enough benefits for

---

[6] Lori's War, *Foreign Policy*, March 2000, p. 28.

most of the population to control political restiveness in a country with many acute ethnic and regional tensions. The great majority of the Indonesian people seemed either intimidated or willing to tolerate the country's failure to live up to the democratic ideal. But when the economy found itself in serious trouble during the last months of 1997, President Suharto had little to fall back upon internally or externally to maintain the political allegiance of the citizenry and his political edifice, which had seemed so formidable just months earlier. The Jakarta regime rapidly crumbled around him. The latent political illegitimacy of the Java-centric Indonesian government became a destabilizing factor that accompanied and intensified the economic and ethnic tribulations of the country.

The receptivity of the political elite to a GPA

Portions of the corporate elite might be persuaded that it is in their interest to support a GPA. Would not those who control state power, however, be less likely to go along with such an innovation? Surely any public institution that could reduce the global democratic deficit by claiming to speak directly for global society could eventually become an important counterweight to state and market power. The important word here is eventually. A relatively weak assembly constituted initially mainly with advisory powers would begin to address concerns about the democratic deficit while posing only a long-term threat to the citadels of state power. This being the case national leaders, whose concerns tend to be associated with short-term prerogatives, have little reason to feel significantly challenged by the establishment of such an assembly. Systemic transformation of world order that could affect successors would not to be threatening to, and might in fact appeal to those political leaders who are themselves most inclined to extend democratic ideals to all arenas of authority and decision.

Putting in place a minimally empowered, but politically saleable institutional structure that nonetheless has far-reaching transformative potential is, in fact, an approach often adopted by the most effective advocates of new global institutions. What has become the European Union, for example, began after the Second World War as the European Coal and Steel Community, a modest, skeletal framework for what would decades later evolve into an integrated European political structure that more recently poses some serious challenges to the primacy of the European state. The French Declaration of 9 May 1950 initially proposing the European Coal and Steel Community makes clear that this humble beginning was by design:

> Europe will not be made all at once, or according to a single plan. [The French Government] proposes that Franco-German production of coal

and steel as a whole be placed under a common High Authority, within the framework of an organization open to the participation of the other countries of Europe. The pooling of coal and steel production should immediately provide for the setting up of common foundations for economic development as a first step in the federation of Europe, and will change the destinies of those regions which have long been devoted to the manufacture of munitions of war, of which they have been the most constant of victims.[7]

Within the European Union, by far the best model for a globally representative assembly, the European Parliament started life as an institutional vessel largely devoid of formal powers. Through time, as the sole direct representative of the European citizenry, the Parliament began to acquire an important institutional role that has given vitality to the undertaking, as well as increasingly reinforcing the European will to carry on with their bold experiment in regional governance.

One source of optimism that many national leaders can be persuaded to support this assembly project arises from the recent experience of building a coalition to push for the establishment of a permanent International Criminal Court. A large number of civil society organizations, working in collaboration with governments, have been very effective, at least so far,[8] in building widespread cooperation among political elites around the world on behalf of a project that only a decade earlier had been dismissed as utopian. The willingness of political leaders to support the creation of such a tribunal is quite surprising. It also lends indirect encouragement to efforts to establish a GPA because the criminal court compromises traditional sovereign prerogatives far more than would be the case initially if a global parliament comes into existence. The court has the substantive power to prosecute individuals for their failure to comply with international criminal law, which means that states have lost exclusive control over the application of penal law, which had been regarded as one of the traditional and fundamental attributes of sovereignty. Government leaders have lost their immunity to some extent in relation to international standards. By comparison a parliament with largely advisory powers would appear to be a relatively modest concession to the growing demand for a more democratic and legitimate global order, and would initially not significantly impinge upon the exercise of sovereign

---

[7] See http://europa.eu.int/comm/dg10/publications/brochures/docu/50ans/decl_en.html#-declaratio

[8] The Statute for the International Criminal Court was overwhelmingly adopted by a conference of states in Rome on 17 July 1998. The Statute received the necessary ratifications and came into force in 2002 despite obstruction from the United States.

powers of a state. Of course, the idea of a parallel international law-making body, even if advisory, does raise the possibility in the moral and political imagination, that more centralization of authority is necessary and desirable, and this possibility, however remote, is likely to be threatening to governments administering nation-states.

Realizing the vision

While the rationale for establishing such an assembly definitely exists, this is, of course, not enough. There needs to be some viable way for this potential to be realized. We believe the formula with the best ability to take advantage of the political promise we have identified can be found in what is being called the 'New Diplomacy'. Unlike traditional diplomacy, which is solely conducted among states, the New Diplomacy is based on the collaboration of civil society with whatever states are receptive, allowing the formation of flexible and innovative coalitions that shift from issue to issue and over time. The major success stories of global civil society in the 1990s were produced in this manner including the Global Warming Treaty, the Landmines Convention and the International Criminal Court.

This New Diplomacy (if it is to continue into this new century) is well adapted to meeting the challenge of creating a globally elected assembly. Nevertheless, the seemingly most natural way to bring a new international regime into being, a large-scale multilateral conference, does not appear well suited to this project. Despite the receptivity of some political elites, there is unlikely to be a critical mass of states in the UN General Assembly or outside its confines that would be willing to call for the convocation of such a conference. We believe that the momentum that would lead to significant state support for the assembly would undoubtedly have to be developed indirectly and gradually. Two other possible approaches seem worth considering in relation to bringing the GPA into being.

One approach that we discuss in more detail in the Summer 2000 edition of the *Stanford Journal of International Law*[9] would be for civil society with the help of receptive states to proceed to create the assembly without resorting to a formal treaty process. Under this approach the assembly would not be formally sanctioned by the collectivity of states and hence its legitimacy would probably be contested by governments at the outset unless they chose

---

[9] See Richard Falk and Andrew Strauss, 'On the Creation of a Global Peoples' Assembly: Legitimacy and the Power of Popular Sovereignty', *Stanford Journal of International Law*.

to ignore its existence altogether. This opposition could be neutralized to some extent by widespread grassroots and media endorsement, and by the citizenry as expressed through popular elections that were taken seriously by large numbers of people and were fairly administered.

The other approach is to rely on a treaty, but to utilize what is often called the Single Negotiating Text Method as the process for coming to an agreement on the specifics of an assembly among supportive states. Pursuant to this approach after extensive consultations with sympathetic parties from civil society, business and nation-states, an organizing committee would generate the text of a treaty establishing an assembly that could serve as the basis for negotiations. Momentum could be generated as civil society organized a public relations campaign and some states were persuaded (sometimes as a result of agreed upon modifications in the draft) to accede to the treaty one at a time. As in the Ottawa Process that ultimately led to the Landmines Convention, a small core group of supportive states could lead the way. Unlike the Landmines treaty, however, which it was thought could not meaningfully come into effect before forty countries ratified it, a relatively small number of countries, say twenty, could provide the founding basis to bring such an assembly into being. Though this number is but a fraction of what would eventually be needed if the assembly wished to have some claim to global democratic legitimacy, it is worth remembering that the European Coal and Steel Community, which evolved to become the European Union, started with only six countries. After all, once the assembly was established and functioning in an impressive way the task of gaining additional state members should become easier. There would then exist a concrete organization to which states could actually be urged to join by their own citizens. As more states joined, pressure on the remaining states to allow their citizens to vote and participate would likely grow, especially if the assembly built a positive reputation in its early years. Holdout states would increasingly find themselves in the embarrassing position of being in a dwindling minority of states denying their citizens the ability to participate along with persons from foreign countries in the world's only globally elected body. It would seem increasingly perverse to proclaim democratic values at home but resist democratic practices and possibilities abroad. The exact nature of the representative parliamentary structure that should or will be created remains to be determined, and should be resolved through vigorous discussion by many different actors drawn from all corners of the world. What is clear to us, however, is that the ongoing phenomena of global democratization are part of an evolutionary social process that will persist, and intensify. While it is still too early to determine the long-term implications of the events of September 11, the future will surely find many ways to remind the peoples of the world that a commitment to global democratic governance is a matter of urgency, and that a way to move forward is through the establishment of a GPA.

Until the onset of the global terror challenge, the two dominant themes of the post-Cold War years were globalization and democracy. Proclamations are now commonplace that the world is rapidly creating an integrated global political economy and that national governments that are not freely elected lack political legitimacy. In view of this, it is paradoxical that there has not yet been a serious global debate on concrete proposals to resolve the obvious contradiction between a professed commitment to democracy at the level of the sovereign state and a manifestly undemocratic global political-economic order. Perhaps this apparent tension can be explained as a form of political inertia, and possibly by the residual sense that such democratizing proposals are still per se utopian. Whatever the explanation, this contradiction will not be tolerated for long. Citizen groups and business and financial elites are not waiting around for governments to come up with solutions. They have taken direct and concrete action to realize their aspirations. These initiatives have created an autonomous dynamic resulting in spontaneous forms of global democratization. As this process continues in an attempt to keep pace with globalization, as it surely will, the movement for a coherent and legitimate system of global democracy will and should intensify. To political elites it will continue to become increasingly obvious that without legitimating institutions, governing the global order will be more difficult and contentious. They are likely to be plagued by the growing disinclination of citizens to accept the policy results of an ever-more encompassing system that is not based on a recognizable form of legitimate governance. To the organized networks of global civil society and business the inclination, reinforced by the practice of democratic societies, is to find direct accommodations and to work out differences. Such a process will naturally lead policy-makers to look toward familiar democratic structures to bridge present, widening cleavages. Finally, to all those who are seriously concerned about social justice, and the creation of a more peaceful global order, the democratic alternative to an inherently authoritarian global system will surely be ever more compelling.

Give Citizens A Voice

by Andrew Strauss and Richard Falk

The Times of India, 2008*

The threat of catastrophic global terrorism is not receding. Most recently, this September in Germany authorities stopped a major terrorist attack against targets which included a US military base and Frankfurt International Airport.

A foiled terror plot against Saudi Arabia's oil industry last spring could have placed the world economy in a tailspin, and Al-Qaida is reconstituting itself in north-west Pakistan at a time the very stability of that nuclear armed nation is in question.

It is time to acknowledge that our efforts to deal with terrorism of global scope are misconceived. An ultimately effective response to those who commit political violence on behalf of widely felt social grievances will require more than law enforcement and military force. We need to do globally what all democracies do: offer those who seek change in civil terms of engagement.

Presently interstate institutions such as the United Nations, the International Monetary Fund and the World Trade Organisation communicate the contrary. They actively exclude citizens from meaningful participation. At intergovernmental bodies, only governments have a seat at the table, and of these, a few dominate.

While individuals with access to political power within the dominant countries can leverage that access into tremendous global influence, average citizens around the world find themselves lacking a global voice. The feelings of alienation that this engenders are all the more acute for those from countries where citizens are also denied domestic participation.

We believe that a singularly meaningful step to introducing global democracy would be the negotiation of a treaty establishing a popularly elected global parliamentary body. It could be modelled on the European parliament, or other new approaches to representation.

* Reprinted from THE TIMES OF INDIA, January 1, 2008.

Enshrined in the founding treaty would be a commitment to universal participation, but mechanisms of implementation would be the subject of negotiation.

Critics in the past have claimed that the establishment of such a parliament is fanciful. But now in a rapidly globalising and highly networked world, no law of political behaviour precludes its creation. National leaders, of course, are always reluctant to give up power and authoritarian regimes would surely reject the project.

But if the parliament was initially proposed as a largely advisory body whose original launch did not require the participation of most countries, the symbolic sacrifice of power by responsible leaders could be outweighed by their commitment to democratic values and to countering terrorism.

Over time, such a parliament could have a significant effect on global politics. The parliament would become a global public space where groups of all persuasion could compete for the moral legitimacy conferred by the parliament's democratic imprimatur.

Perhaps it might, along with the UN General Assembly, eventually become part of a bicameral global legislative system. If our commitment to the democratic spirit is to be more than rhetorical, we must begin to democratise the increasingly powerful global system.

Only then can citizen participation in the system become institutionalised and will we have hope of gaining global acceptance for the core democratic norm that renders political violence beyond the pale. It would be naive to expect fundamentalists of the sort who belong to Al-Qaida to accept an offer to foreswear political violence in exchange for democratic access to the global system.

But long experience in democratic societies shows that when democracy becomes normalised those who employ violence find themselves increasingly isolated and ineffectual.

This reasoning may seem far from our present political reality, but recently a global campaign for a UN parliamentary assembly was launched with the support of almost 400 parliamentarians, 20 acting and former national government ministers including two former prime ministers and six former foreign ministers, four noble laureates and a former United Nations secretary general.

In these troubled times people are finally looking with favour on global democratic alternatives to our violence-prone world.

6
On the GPA as a Practical Political Project

Not A Parliament of Dreams: A Response to Joseph Nye

by Richard Falk and Andrew Strauss

World Link, 2002[*]

T he way the world is run is not sufficiently democratic. On that much we agree with Joseph Nye, the dean of Harvard University's Kennedy school of government. As he explained in "Parliament of Dreams" (World Link, March/April 2002), international organisations — which are assuming new responsibilities because "many of the issues raised by globalisation are inherently multilateral" — fall short of the democratic standards that we expect elsewhere.

But Mr Nyc's response to this "democratic deficit" is inadequate. He proposes vague initiatives such as "greater clarity about democracy, a richer understanding of accountability and a willingness to experiment." Disappointingly, he disparages our suggestion that this crisis of democratic legitimacy requires the creation of an elected global parliamentary assembly (GPA), modelled on the European Parliament.

Our proposal could transform the way the world is run — yet it is politically realistic. The eventual goal is a world body with limited but important legislative powers that would enfranchise adult citizens everywhere. This ambitious aim is realistic because it could be achieved incrementally. Even 20 to 30 pioneering countries that are geographically and economically diverse enough to be credible founders could launch the GPA. Its powers could initially be merely advisory, to avoid intimidating national leaders. If civil-society organisations rallied behind it, as they did with the Ottawa Landmines Treaty or the International Criminal Court, a GPA could see the light of day.

The second half of the 20th century has taught us that grand plans to transform global governance at a stroke are doomed to failure. Perhaps most well-known is the "world peace through world law" treatise. Louis Sohn's

[*] Reprinted from WORLD LINK, THE MAGAZINE OF THE WORLD ECONOMIC FORUM, July/August 2002.

and Grenville Clark's proposal to create a limited world government by dramatically strengthening the United Nations never came close to getting off the ground. In contrast, the stop-and-start — but robust — incrementalism of European integration has proved a big success. Although the architects of a united Europe envisioned a comprehensive European system from the start, what is now the 15-strong European Union started off as the six-member European Coal and Steel Community. The European Parliament, which is now directly elected and can veto around four-fifths of EU legislation, started off as a glorified advisory body.

Once in place, a GPA would, we believe, grow in influence and stature. Its election would give it a visibility unique among international organisations — and its exceptional claim to popular legitimacy would encourage citizen groups seeking to promote their causes to petition it to pass supportive resolutions. Opponents of those resolutions would doubtless lobby too — and thus would begin the familiar process of parliamentary politics writ large. The GPA would become a much-needed venue for global interest groups to interact without having to rely on national governments as intermediaries.

Once the GPA was up and running, citizen groups everywhere would doubtless petition their governments to join the assembly. When a critical mass of membership was reached, even authoritarian governments would find it hard to deny their citizens the right to be represented through free and fair elections in the only global democratic institution. As the GPA evolved, its legal powers, and its relationship with the UN, would have to be agreed as part of a wider effort to establish effective and equitable global governance. Perhaps, along with the UN General Assembly, it could form the basis for a two-chamber global legislature. The ultimate aim is not to create a world state that apes sovereign nation states. It is to build a form of global governance that is both effective and democratic: capable of tackling regional and global problems and responsive to people's views.

It might take years — if not decades — to get there. But the GPA's mere existence would give it a moral influence equal to, if not exceeding, that of existing NGOs. By holding hearings, issuing reports and passing resolutions, even a fledgling GPA could make other global bodies more accountable.

Mr Nye disagrees. He argues that a world parliament is not practical because "minorities acquiesce in the will of the majority when they feel they participate in a larger community" and "there is little evidence that a sufficiently strong sense of community exists at the global level or that it could soon be created". True, a strong sense of global community seems to be lacking, although that could also be said of many countries. But even if his assessment of current attitudes is correct, the parliamentary structure we propose does not require a mature global community.

Mr Nye's emphasis on the importance of community in parliamentary decision-making seems based on a model of Athenian democracy suitable for ancient city-states. It is hardly adapted to the modern era of mass politics. Day to day, the successful functioning of modern pluralist democracies depends primarily on organised interests and pressure groups — not minorities within the unorganised (and largely politically dormant) community as a whole — accepting policy decisions. For example, even when a vote goes against them, environmentalists and the industries they spar with continue to support the parliamentary process because they believe in it and value their ability to participate in it. If Mr Nye's reasoning was correct, the International Criminal Court — which is in some ways more radical and more threatening to world leaders than a GPA — would never have been established. Besides, a GPA would help nurture a sense of global community. Contrary to what Mr Nye assumes, political communities are often formed by institutions rather than preceding them.

To illustrate his point, Mr Nye argues that "treating the world as one global constituency implies the existence of a political community in which citizens of around 200 states would be willing to be continually outvoted by more than a billion Chinese and a billion Indians." This, he claims, would be a "nightmare for many of the protesting NGOs that seek to promote international environmental and labour standards, as well as democracy". Yet in existing parliaments that represent multinational constituencies — India's, Belgium's or the European Parliament, for instance — members generally vote on the basis of substantive concerns and only rarely along ethnic lines. Unlike government representatives at international institutions, Indian and Chinese delegates to a freely elected global assembly are unlikely to adopt a unified position. They would probably reflect the diversity of interests and outlooks within their societies, including on issues such as labour and environmental standards. Shifting transnational coalitions would expose the fiction of coherent national interests. They might also help to mitigate dangerous conflicts among governments through parliamentary operations that evolve a political culture less beholden to nationalism and more engaged in promoting human security.

Mr Nye's most serious and most perplexing criticism is that popular elections "might well produce an undemocratic body that would interfere with the delegated accountability that now links institutions to democracy". How could a democratically elected body possibly interfere with global democracy? His reference to delegated accountability suggests that the democratic credentials of today's international bodies come from the elected national governments that are represented there. But as he recognizes elsewhere in his article, this reliance on government representation gives the international system at best an attenuated claim to democratic legitimacy.

Far too many national leaders are not democratically elected, yet the current international system — unlike a GPA — accords them unconditional authority to "represent" their countries. Even elected national leaders, though, should not have unchecked powers to decide global policy. In the absence of a collective democratic structure such as a GPA, international decision-making is an institutionalized process whereby states use their leverage to pressure other states. This enables political elites in powerful countries to impose their preferences on foreign citizens. Popular resentment towards this coercion and lack of representation helps explain growing grassroots frustration and much of the anger directed at the almighty US.

The core of our disagreement with Mr Nye stems from our differing views of the international system. He seems to regard as acceptable, or perhaps inevitable, the current international system, whose fundamental organizing principle is state sovereignty, not citizen democracy. He seems concerned only with how the system can be tinkered with to quell the rage of the anti-globalization movement.

We see this popular outcry as a valuable political opportunity to launch a process of fundamental transformation. The current system based on state sovereignty is increasingly authoritarian and dysfunctional, tends towards recurrent war and allows governments to disregard international law and morality, which are of common interest. We hope and believe that a global parliamentary assembly could help pave the way to an international system where social questions and political conflicts are settled in as peaceful and fair a way as they are in the world's more successful democracies.

Letter to The Editor in Response to Joseph Nye

by Richard Falk and Andrew Strauss

Foreign Affairs, 2001[*]

Joseph Nye argues against the creation of a global parliamentary assembly and instead proposes several more modest initiatives to remedy the global democratic deficit ("Globalization's Democratic Deficit", July/August 2001). He suggests that a parliamentary body would be unworkable because "in a democratic system, minorities acquiesce to the will of the majority when they feel they are generally full-fledged participants in the larger community" and because "there is little evidence ... that such a strong sense of community exists at the global level today, or that it could soon be created." If Nye means that conditions are not appropriate for a global version of the U.S. Congress, then he surely is correct. But this consideration does not refute an argument in favor of establishing a global advisory assembly — at the initiative of willing states or civil society. Such an assembly could monitor global institutions, and, like the European Parliament, it might well enhance its reputation and strengthen its authority over time.

The successful day-to-day functioning of modern democracy depends not as much on the acceptance of policy by factions within the community as on such acceptance by organized interests. For example, those individuals actively supporting environmental groups tend to continue their fundamental allegiance to the parliamentary process despite losses to industry because they are committed to the validity of the system and know that they will have the opportunity to press their case on another day. In fact, if influential interest groups that are directly involved in policymaking at the domestic level can circumvent states and directly interact through the vehicle of a global parliament, support for and adherence to the international system is likely to be strengthened significantly.

[*] Reprinted by permission of FOREIGN AFFAIRS, Volume 80, No. 5, September/October 2001. Copyright (2001) by the Council on Foreign Relations, Inc.

Nye suggests that "treating the world as a single global constituency ... would mean that more than 2 billion Chinese and Indians could usually get their way." Indian and Chinese delegates to a freely elected global assembly seem unlikely to adopt a unified position, and if they did, they would discredit themselves and weaken the assembly. Rather than form a unified bloc, they would more likely reflect the diversity of interests and outlooks within their respective societies. Shifting transnational coalitions would expose the fiction of "national interests" and help mitigate dangerous conflicts by way of parliamentary politics.

Letter to The Editor in Response to Anders B. Johnsson

by Richard Falk and Andrew Strauss

The Nation, 2003[*]

Anders Johnsson contends that our proposal for a global parliament is unrealistic because many countries do not yet have authentic national democracies. He assumes that democracy can proceed only in one specifically prescribed linear fashion, from local to global.

We do not agree.

In the days when many of America's biggest cities were still run by anti-democratic political machines, no one would have claimed that the United States should abandon democracy at the national level because, as Johnsson asserts, it always starts at home.

In fact, beginning to establish democratic practices globally promotes democratization locally. If there existed a global parliament whose membership was restricted to freely elected parliamentarians, authoritarian leaders would be under pressure to choose: Either hold fast to the politically embarrassing option of denying their people the right to be represented in the world's only globally elected body or allow the introduction of democratic practices along with the emergence of a democratically legitimized leadership. As globalization proceeds and as decision-making power is increasingly situated at the international level, Johnsson asks us to wait indefinitely for global democracy until all countries are democratic. But for those of us in more democratic countries, to wait is to watch as our own hard-won democratic space becomes more and more restricted.

Johnsson is right, though, to put an emphasis on being realistic. Too often in the past, discussions of fundamental world-order reform have been too easily dismissed as utopianism. But, what is, after all, so unrealistic about our proposal? Is it unrealistic to suppose, as we suggest, that twenty or thirty more enlightened governments might agree to a treaty creating a standalone advisory parliament with very limited initial legal powers? This is, after all,

---

[*] Reprinted from THE NATION. November 3, 2003.

the way the European Parliament began. Is it unrealistic to think that such an assembly, as the first popularly elected global body, would attract considerable attention and that various interests would attempt to gain the support of this singular democratic body to legitimize their policy positions? And, is it unrealistic to think that such a body would become a focal point for the pursuit of global reform, leading to expanding influence, inducing increases in membership and powers?

Of course, Johnsson is right that there are complex logistical, representational and jurisdictional matters to resolve in establishing such an international body. But, these matters present themselves whenever a new political organization is brought to life, and, yes, they are challenging, but far from impossible to overcome. We are, after all, living in a period when little over a decade ago the Soviet Union ceased to exist and the establishment of the International Criminal Court seemed a distant dream.

As for John Kelly's proposal to move toward a more democratic global order by electing representatives to the UN, we find the idea worthy of further discussion; but we would question its feasibility, as it would mean displacing the representational role of governments altogether at the UN. We offer our own proposal not out of a conviction that it is the only way to proceed but from a sense that we need to begin a serious global discussion about how the world order can be made more democratic, peaceful, fair and sustainable.

Reviving the Dream of Global Democracy

by Richard Falk and Andrew Strauss

in Hope in A Dark Time (Krieger ed.), 2003[*]

The 1990s brought great hope to humanity: the peaceful end to the Cold War and to apartheid, the liberation of East Europe, sustained economic growth in many of the poorest regions of the world, steps toward holding political leaders responsible for crimes of state, the spread of democracy and human rights, the strengthening of regional institutions, especially in Europe, the emergence of global civil society as a real political force, the impact of social action designed to make globalization more equitable and environmentally protective, and robust coalitions of governments and civil society actors that completed projects such as a treaty prohibiting anti-personnel landmines and an agreement to establish an international criminal court. Without exaggeration, a normative revolution was underway that if sustained, would result in "a new world order" based on humane global governance, and of benefit to all peoples on the planet.

Of course, all was not well. There was much violence and turmoil in the world, and bloody civil wars raged in several parts of sub-Saharan Africa and the Balkans. The United States did not exert leadership in world affairs during the years following the collapse of the Soviet Union to move toward nuclear disarmament or to establish an independent peacekeeping force under the authority of the United Nations. And "the peace process" that was supposed to end the Israel/Palestine encounter turned out to be a failure: Globalization also ran into some speed bumps in the form of the Asian Financial Crisis, as well as a rising tide of protest and a series of emergencies arising from countries such as Turkey, Argentina, and Indonesia on the brink of bankruptcy and chaos.

And yet, overall, these problems seemed manageable with wisdom and patience.

Then came the terror attacks of September 11, and an American response that opted for global war, starting in Afghanistan, but dedicated to a long

---

[*] Reprinted from HOPE IN A DARK TIME (David Krieger, ed.) (Capra, 2003).

struggle on many fronts with the visionary goal of eliminating the terrorist dimension from international life. It has seemed that the power of the United States was such that it could reshape the global agenda overnight, making its campaign against terrorism dominate all other issues, and as in the Cold War, put on indefinite hold projects for global reform. Whether such control will be successfully maintained is difficult to discern at this point, but it will depend on the degree to which citizens associations and grassroots initiatives rise to the challenge, and recover the momentum that had been gathering so impressively in the 1990s. There are some early encouraging Signs: it appears that an International Criminal Court will exist before the end of 2002, the World Social Forum in Porto Allegre, Brazil was the largest meeting ever of citizens from around the world seeking an alternative globalization, and there appears to be growing support for the establishment of a fully sovereign Palestinian state that shares Jerusalem as its capital.

A Global Peoples' Assembly

In this essay, we explore another test case: whether the social forces around the world can carry forward the vision of global democracy, and more specifically, whether a start can be made during this decade on a vital component of this vision, the creation of a Global Peoples' Assembly (GPA) or parliament. It seems essential to have such an institutional presence for the expression of ideas, grievances, and proposals that are not acceptable to leading governments, but are needed for the wellbeing of the peoples of the world. The practicality of this initiative has been demonstrated in Europe where the European Parliament has gradually and fitfully grown from a neglected talk shop to becoming an integral dimension of European political life, and a key counterweight to concerns that popular sentiments and societal values were being stifled by the Eurocrats in Brussels. The European experience provides a model of constitutional structure and of process, whereby an innovative institution learns, adapts, and continuously reinvents itself.

The justification for global democracy relates to the new ways in which global policy is being made, and the degree to which nonterritorial networks are undermining the authority and control of territorial actors. The dark side of networking is evident in relation to terrorism, drugs, crime, illegal migration, financial speculation, and environmental decay. But networking has its brighter sides, as well, with the Internet and information technologies generally, providing peoples with extraordinary instruments of empowerment. It enlarges the possibilities of participation, awareness, and accountability by many orders of magnitude. In fundamental respects, the push for global democracy is to give political shape to these developments, both to resist the dangers associated with globalization-from-above, including unbridled corporate and financial control of the world economy and the prospects of the militarization of space, as well as to move ahead with the opportunities associated with globalization-from-below, especially reviving the normative revolution of the 1990s. The aspirations for global democracy will only remain credible if concrete steps are taken to give these Widely shared sentiments some sort of institutional presence. For this reason alone the project to establish a GPA in the coming years deserves a high priority from activists around the world.

Such a priority is also needed to overcome some resistance to democratizing moves that had been earlier underway. One of the exciting developments during the 1990s was the extent to which representatives of civil society were active participants at global conferences on big issues such as environment, human rights, women's issues, and social development that worked in collaboration with more progressive governments to exert influence on the positions taken by the assembled governments. The very success of this informal experiment in global democracy produced a kind of backlash, leading some states, particularly Russia, China and the United States, to mount criticism of such events as wastes of time and money, mere "spectacles." As a result, the UN can no longer integrate civil society actors into its main consciousness-raising activities.

New Forms Of Participation Are Needed

It was a notable additional contribution of the United Nations to give credibility to the idea that the Organization needed to incorporate into its activities both representatives of civil society and of the private sector. Kofi Annan, during his distinguished tenure as Secretary General, has frequently given eloquent expression to this view that governmental representation is no longer enough to ensure the continued relevance of the United Nations. In a world of transnational activism and networking, new forms of participation must be found. The Secretary General took the lead, despite some resistance from prominent member states, in sponsoring a Millennial Assembly of Civil Society on a one-time basis to make a contribution to the various observances of the year 2000.

It is an open question of feasibility and effectiveness whether to work for the establishment of a GPA as a second assembly within the framework of the United Nations or as a free-standing institutional presence. The advantage of the latter approach is that it can begin to happen as soon as enough funding and political support is forthcoming. There have been various recognitions of the relevance of the GPA idea that can be viewed as seedlings. Steps have been taken by individuals and groups in several coun-tries to establish shortly an e-parliament that would engage in debate and produce proposals entirely online. Another creative development has been the biennial Assembly of the United Nations of Peoples held in Perugia, Italy on four separate occasions, bringing together representatives from over 100 countries to discuss common concerns over a period of several days. The Perugia events, as does the Porto Allegre Social Development Forum, show clearly that the voices of the peoples of the world are giving expression to a range of different demands, dreams, and proposals that come to the surface in inter-governmental organizations.

Against this background of thought and action, it seems possible to high-light this project to establish a functioning GPA within the next several years. There are many obstacles and challenges that make the task seem for-midable, but as has been often suggested in such circumstances, even the longest journey begins with the first steps. It will be important to take these steps to rekindle the hopes of peoples around the world that the September 11 attacks not be allowed to put on hold the various dimensions of global reform, and more specifically, the importance of globalizing democratic practices and procedures in a manner that takes account of globalizing trends with respect to the organization of the world economy and international secu-rity.

Military globalization, led by the United States, with its plans for missile defense and the weaponization of space, its distrust of arms control arrange-ments, its unilateralism, makes this effort to build an effective global demo-cratic presence of urgency. We can expect the poor and most vulnerable peo-ples on the planet to be particularly threatened by this prospect of American ambitions to control the earth as a whole on the basis of its dominance in military might. A consensus for demilitarization, for nuclear disarmament, for peace and justice needs to be shaped to offer alternative concepts of hu-man security and of a world dedicated to the use of its resources to meet the needs of people rather than the greed of markets and the geopolitical ambi-tions of rich and powerful states.

Of course, the GPA need not be, and likely would not be, antistatist, alt-hough it would almost certainly be critical of dominant economic and securi-ty policies being currently pursued by leading states, and of the misallocation of resources to bolster the military. Indeed, it would likely build on the im-pressive collaborative experience in the 1990s when coalitions of Non-Governmental Organizations (NGOs) joined with like-minded governments to promote a common agenda that was seeking to get policy results in rela-tion to war/peace, environment, and human rights concerns.

The Path To A Global Peoples' Assembly

The choice of method for the pursuit of the project of a GPA is of great tactical relevance at this stage. The most conventional approach would be to form a coalition of NGOs around the world that would seek to influence governments to hold a conference at which a treaty of establishment would be tabled for signature and subsequent ratification. The treaty could specify that the GPA would start to function when 30 governments had deposited an instrument of ratification with the UN, and had contributed an assessed amount need for a startup budget. An alternative approach would be to seek voluntary contributions, a founding civil society convention perhaps linked to a meeting of the World Social Forum, and an informal process of establishment based on annual sessions of the GPA held in various parts of the world, or located in either Geneva or New York for maximum impact on the UN system and the world media.

Whatever path is pursued, the founders could delimit the functions and powers of the GPA. If established outside the UN and without the formal participation of states, then it would emerge as a project of global civil society, which would take on the task of building the structures for global democracy. If states and the UN participated, then the GPA would operate within the formal framework of world order, and its innovative status would be viewed as a less radical type of global reform. In either event, the growth of influence and legitimacy would depend on the support that the GPA could gain for its early efforts, and how it conceived of its role, whether as a citizens' dimension of the existing structure or as a transformative platform from which to work for humane global governance, including nuclear disarmament, war prevention, and a comprehensive functioning rule of law.

In any event, the movement for the establishment of a GPA would link peoples throughout the world in efforts to establish the key institutional component of an emergent global democracy. The process would itself be a manifestation of the democratic spirit, a learning experience of political value, and a contribution to defining and supporting the idea of "global democracy." As such, it would provide activists everywhere with a hopeful focus for their grievances, and a practical means by which to work for peaceful change and social justice in the difficult years ahead.

For a Global Peoples Assembly

by Richard Falk and Andrew Strauss

The International Herald Tribune, 1997[*]

The recent dramatic announcements of record-setting contributions to international causes by Ted Turner and George Soros suggest tremendous possibilities for the future.

These two men signify the rise of a new breed of global philanthropist active in fashioning an international civil society. It was globalization that gave them the opportunities to amass extraordinary wealth. It now provides them and others with a unique opportunity to contribute to human well-being.

This includes pushing for the democratization of the global order, a goal that governments are reluctant to promote.

Such individuals could do this most imaginatively by providing funds for the establishment of a popularly elected Global Peoples' Assembly, which would provide the world's citizens for the first time with a forum to express their planetary aspirations and grievances outside the traditional nation-state context.

Elections for this assembly could be organized and administered by an international citizens' committee and overseen by the respected Swedish organization International Democratic Elections Assistance, or IDEA. Once established, the assembly could lobby governments for formal recognition within the UN system.

To begin with, however, such an assembly would have an international legal status similar to that of nongovernmental organizations like the Red Cross or Amnesty International. Unlike them, however, it could lay claim to speak on behalf of the peoples of the world. As the only such body, it would have the potential to be highly influential even before receiving formal recognition.

Specifically, how would this assembly make its influence felt? Like the UN General Assembly, whose official powers are largely recommendatory, such an assembly would contribute to the creation of planetary norms of behavior by issuing resolutions and proclamations, and more generally by expressing views on critical issues of global policy.

In a more and more integrated world that increasingly ascribes to democratic principles, the case for such an assembly seems unassailable.

First, because the globalization of the world economy inevitably requires the development of global regulatory institutions, the preservation of freedoms now enjoyed demands we begin to structure these institutions along democratic lines.

Second, the very existence of a citizen-controlled international assembly would both ideologically and practically reinforce democratic practices within countries and undermine authoritarianism.

Third, allowing representatives from different countries and civilizations to work together to advance mutual interests and discuss differences in an assembly setting would help promote a climate of civility in global affairs, encouraging universal values to prevail over more parochial concerns, as well as over sectarian loyalties and beliefs.

Finally, the establishment of such a global assembly with direct electoral accountability to workers, peasants and other citizens would give currently vulnerable groups a voice and help them regain some of the power lost to international capital as a result of globalization.

The major argument likely to be advanced against such an undertaking is that it is naive, idealistic and, at best, premature. To be sure, logistical problems would have to be overcome. Worldwide elections would have to be independently organized. A voting formula based upon one person, one vote would have to be put into place, and elections would need to be certifiable as free and fair.

There would, of course, be glitches. Some governments would undoubtedly not allow such elections to occur on their territories, and until sufficient pressure could be brought to bear their citizens would have to go unrepresented. But these problems would not be fatal to the endeavor.

There is no reason to think this lies beyond the realm of the possible. Indeed, a bold, visionary undertaking at the start of a new millennium might activate the political and moral imagination of all those who aspire to construct a world order more responsive to the values associated with democracy.

Those with the resources have the capacity to make this proposal a reality by seizing the initiative and promoting the democratization of the emerging international order. Democracy at the global level is needed and long overdue.

All That Dough

by Andrew Strauss and Richard Falk

The Philadelphia Inquirer, 1997[*]

Global media giant Ted Turner's dramatic announcement that he intends to give a billion dollars to the United Nations over the next 10 years suggests tremendous possibilities. He and international financier George Soros signify the rise of a new breed of global philantropists. Globalization gave such people opportunities to amass extraordinary wealth; now it gives them an historic opportunity to contribute to the democratization of the global order.

Their money could help establish a popularly elected Global Peoples' Assembly which would for the first time give citizens of the world a forum outside the traditional nation-state context. Elections for this assembly could be run by an international citizens committee. Once in place and operative, the assembly could proceed to lobby governments for formal recognition within the United Nations system.

The very existence of a citizen-controlled international assembly would both reinforce democratic practices within countries and undermine authoritarian rulers. As representatives from different countries and civilizations work together, a climate of civility in global affairs would arise.

Democracy should extend beyond national borders. Why? The more we globalize the world economy the more we need international regulation of planetary social life. If a global polity is inevitable, then — to preserve freedoms — we should design that structure along democratic lines.

Naive? Idealistic? Premature? That is what critics would say. But if wealthy cosmopolitans such as Turner and Soros could agree on the benefits of providing seed money it could happen. And a bold, visionary undertaking at the start of a new millennium might excite the imaginations of all those who aspire to a world more responsive to the values associated with democracy.

[*] Reprinted from THE PHILADELPHIA INQUIRER, October 12, 1997.

For the first time, we have the conditions for a global democratic system, in which respect for human beings is not limited by national borders, and in which governmental power, domestic and international, is derived directly from the consent of the governed. Those with the resources can help make this proposal a reality. Global democracy is long overdue.

7
Developments Towards Global Democracy in a Changing World Order

What Comes After Westphalia:
The Democratic Challenge

by Richard Falk

Widener Law Review, 2007*

There exists a puzzling disconnect between the almost universal advocacy of democracy as the sole legitimate way to organize domestic society and intense resistance from leading state actors to any steps taken to democratize the ways in which global governance in its present form is constituted and administered. There exists a particularly striking contrast between the political language that has been used by the current American political leadership in the course of the Bush presidency, which has made its signature claim to moral leadership in the world depend on its supposed championship of democracy while at the same time displaying an active hostility toward democracy as it might inform global governance. The neoconservative version of this disconnect is more explicit than a similar "democratic gap" that existed earlier, and was especially characteristic of the Clinton presidency, which also supported the spread of democracy on the national level as an essential element of its foreign policy (what it called "enlargement"). As with Bush, Clinton also was unsupportive of civil society's efforts to open up the United Nations, or global governance more generally, to the impact of democratizing pressures. An inquiry into global democracy proceeds against this background of understanding.

The idea of global governance is itself elusive. It is a term of art that has come into being rather recently, at least most prominently, to consider the need for and form of governmental capabilities at the global level without implying the existence or desirability of world government.[1] There is considerable sensitivity on this matter of language as "world government" is asso-

---

* Reprinted from WIDENER LAW REVIEW, Volume 13, Issue 2, 2007.

[1] COMM'N ON GLOBAL GOVERNANCE, OUR GLOBAL NEIGHBORHOOD: THE REPORT OF THE COMMISSION ON GLOBAL GOVERNANCE (1995); *see also* GOVERNANCE WITHOUT GOVERNMENT: ORDER AND CHANGE IN WORLD POLITICS (James N. Rosenau & Ernst Otto Czempiel eds., 2003) (1992).

ciated with the movement for "world federalism," which in turn is derided as utopian or as likely to pave the way toward tyranny on a global scale.[2] The idea of global governance, in contrast, is firmly situated in most formulations at the interface between realism and liberalism, grounded in the resilience of Westphalian world order based on the interplay of sovereign states and on the liberal effort to promote international cooperation and collective action as ways to promote humane values without requiring modifications in the structure of world order.[3] The interest in global governance reflects a growing sense that a stronger set of institutional procedures and practices are needed at the global level to address a series of challenges associated with the global commons, including climate change, polar melting, deforestation, and ocean fisheries.[4] This interest also reflects regulatory concerns about a range of issues, including transnational crime and international business operations. Increasingly, there exists an acknowledged need for a normative framework for economic globalization that will ensure greater poverty reduction and a less unequal distribution of the benefits and burdens of growth on a global scale.[5] Such a preoccupation with global governance can also be thought about as an evolutionary stage in the unfolding of Westphalian world order; in effect, a geopolitical successor to the simpler mechanisms of so-called "Great Power" management of international society that provided all societies with the benefits of global stability, which can be considered as a collective public good.[6] Another way of conceiving of the present historical circumstances is to postulate a "Grotian Moment," that is, a transitional interlude that is signaling a tectonic shift in world order.[7] We are presently expe-

---

[2] For a classic critique of world government as a solution to the crisis of global governance, *see* INIS L. CLAUDE, JR., SWORDS INTO PLOWSHARES: THE PROBLEMS AND PROGRESS OF INTERNATIONAL ORGANIZATION (1971).

[3] For a further exposition and critique, *see* RICHARD A. FALK, THE DECLINING WORLD ORDER: AMERICA'S IMPERIAL GEOPOLITICS (2004).

[4] On climate change, *see* NICHOLAS STERN, THE ECONOMICS OF CLIMATE CHANGE: THE STERN REVIEW (2007); INTERGOVERNMENTAL PANEL ON CLIMATE CHANGE, CLIMATE CHANGE 2007: THE PHYSICAL SCIENCE BASIS (Feb. 5, 2007), *available at* http://ipccwg1.ucar.edu/wg1/-docs/WG1AR4_SPM_Approved_05Feb.pdf (last visited Apr. 3, 2007) (Summary for Policymakers, Contribution of Working Group I to the Fourth Assessment Report of the Intergovernmental Panel on Climate Change).

[5] RICHARD FALK, PREDATORY GLOBALIZATION: A CRITIQUE (1999).

[6] For influential depictions of this theory, *see* HEDLEY BULL, THE ANARCHICAL SOCIETY: A STUDY OF ORDER IN WORLD POLITICS (2d ed. 1995) (1977); *see also* F. H. HINSLEY, POWER AND THE PURSUIT OF PEACE: THEORY AND PRACTICE IN THE HISTORY OF RELATIONS BETWEEN STATES (1963).

[7] For more discussion of the "Grotian Moment" and its surrounding conceptual context, *see* INTERNATIONAL LAW AND WORLD ORDER 126586 (Burns H. Weston, Richard A. Falk, Hilary Charlesworth & Andrew K. Strauss eds., Thomson/West 4th ed. 2006) (1980).

riencing both the terminal phase of the Westphalian framework and the emergence of a different structure of world order that is sufficiently receptive to the emergence of supranational forms of regional and global governance, as well as exhibiting the agency of non-state actors, as to qualify as "post-Westphalian." This assertion, in part, represents recognition that states are incapable of adapting to mounting global scale challenges without a significant reconfiguration of world order. This assessment is not meant to suggest that states have lost their primacy in global political life, but rather to observe that a sustainable world order in the future depends on some major structural and ideational innovations to protect an otherwise endangered global public interest in the years ahead. Institutional and normative expressions of regional and global solidarity will be needed to address such issues as climate change, regulation of the world economy, establishment of security, and implementation of the ethos of a responsibility to protect. Sustainability will also depend on taking into present account the needs of future generations, with respect to resources and the foundations of life supportive of individual and collective human dignity.

More than the United Nations or even the Bretton Woods institutions and the World Trade Organization, the extraordinary regionalizing developments in Europe over the course of the last half century prefigure a post-Westphalian world order that draws on a number of complementary structural and attitudinal ideas to solve the deepening crisis of global governance. The European Union (EU) can be conceived as foreshadowing such modifications on the regional level in Europe, and potentially elsewhere, in a manner that seems entirely consistent with democratic values and procedures.[8] Europe has achieved internal mobility, a common currency, economic progress, regional governance, limitations on internal sovereignty, and most impressively, a culture of peace that makes intraregional arms races, interstate uses of force, and wars almost unthinkable. In current debates about the future of Kosovo, it is being influentially claimed that the only serious hope for reconciling the strong Kosovar push for national independence with the Serbian insistence on the unity of its state boundaries is for both of these contending entities to be formally absorbed into the larger reality of Europe by a new cycle of EU enlargement.[9]

---

[8] For optimistic assessments of the European experiment in regional governance, see MARK LEONARD, WHY EUROPE WILL RUN THE 21ST CENTURY (2005); JEREMY RIFKIN, THE EUROPEAN DREAM: HOW EUROPE'S VISION OF THE FUTURE IS QUIETLY ECLIPSING THE AMERICAN DREAM (2004).

[9] *See* Timothy Garton Ash, *Why Kosovo Should Become the 33rd Member—and Serbia the 34th*, GUARDIAN UNLIMITED, Feb. 15, 2007, http://www.guardian.co.uk /commentisfree/ story/0,,2013344,00.html (last visited March 30, 2007).

It is notable, although ironic, that it is Europe, which invented Westphalian world order back in the seventeenth century, that is taking the lead in
shaping a radical post-Westphalian form of governance for its region. Of
course, Europe manipulated the state system for as long as possible to serve
its geopolitical ambitions, which led to the colonizing of much of the non-
Western world and subordinating most of the rest. In this respect, the EU
should be mainly understood as a belated response to a series of European
geopolitical setbacks as it is an expression of European creativity, or even
less so, European idealism. The anti-colonial movement, the debilitating
impact of the two world wars, the challenge posed by Soviet expansionism
during the Cold War, and the difficulties of competing in the world economy
all played a part in moving European leaders to seek greater unity through
mutually beneficial cooperative practices and procedures. As is well known,
the growth of the EU from its outset was premised on an appeal to the self-
interest of individual sovereign states, especially with respect to economic
policy. It is only by stages that this European experiment in regional world
order began to take shape and a regional political and cultural consciousness
emerged.

Such an understanding helps us realize that normally there are two major
ways of stimulating significant world order reforms: the first, illustrated by
the establishment of the League of Nations and the United Nations, is associated with efforts to reconstruct world order in the aftermath of a destructive
war;[10] the second, best illustrated by the EU, is based on the evolutionary
potential of building upon modest functional beginnings, where the benefits
of institutional growth are weighed periodically by participating governments and their publics, leading to forward surges generally formalized by
treaties negotiated and approved by the EU membership, but also by back-
sliding in periods of disenchantment with aspects of this momentous political
experiment.[11] Since 2005, there has been serious debate about whether the
EU has reached, or possibly even exceeded, prudent limits on its scope (the
enlargement issue) and depth (the question of the European Constitution).
European public opinion has been recently agitated by the costs of enlargement, the tensions associated with immigration, the controversy over possible Turkish membership, and the interplay between Islamic extremism and
Islamophobia. Such incidents as the assassination of Theo Van Gogh, the

---

[10] For an important study of this dynamic, *see* G. JOHN IKENBERRY, AFTER VICTORY: IN-
STITUTIONS, STRATEGIC RETREAT, AND THE REBUILDING OF ORDER AFTER MAJOR WARS
(2001).

[11] For a useful account, *see* DUSAN SIDJANSKI, THE FEDERAL FUTURE OF EUROPE: FROM
THE EUROPEAN COMMUNITY TO THE EUROPEAN UNION (Univ. of Mich. Press 2d ed. 2000)
(1992).

Danish cartoon controversy, the French urban riots, and left views that the EU was anti-worker and a vehicle for neoliberal globalization, were instrumental in the French and Dutch rejection of a proposed European Constitution.[12] Despite this recent cascade of discouraging developments that have certainly cooled some of the enthusiasm about the EU as a model of world order, there remain important reasons to expect a rebound in confidence, as well as to reaffirm this set of regional initiatives to be an extremely positive demonstration that post-Westphalian change and reform is possible to achieve by peaceful means: the European Parliament shows that electoral democracy can be made to work in multistate, multinational political domains; environmentalist pressures to reduce carbon emissions are being most effectively articulated and organized under the auspices of the EU; and along similar lines, the advocacy of a more moderate approach, relying on diplomacy and law rather than force in responding to such threats as posed by political Islam and nonproliferation, is being led by European statesmen.[13] In 2003, opposition of such stalwart American allies as France and Germany to the proposed invasion of Iraq vividly illustrated a growing divergence in approach to world order as between Europe and the United States that especially related to attitudes toward force and war as policy options of governments.

This apparent European submission to the Rule of Law encourages a soft-landing in a post-Westphalian world order. In contrast, the United States, especially during the Bush II presidency, has been far more reliant on a militarist approach in fashioning its efforts to move beyond Westphalian world order, including the seeming acceptance of the inevitability hard landing associated with wars, financing a worldwide network of military bases, and relying on the militarization of space for control over the entire earth.[14] That is, Europe since the end of the Cold War, and especially since the presidency of George W. Bush and the ascent to influence of a neoconservative entourage of political advisors has developed a regional self-consciousness that is defined in part by seeking an alternative path to world order that is less likely to produce catastrophic results. Whether this regional experiment, which can be compared with far less evolved regional frameworks in Latin America,

---

[12] Ian Buruma, Murder in Amsterdam: The Death of Theo van Gogh and the Limits of Tolerance (2006).

[13] *See* Robert Kagan, Of Paradise and Power: America and Europe in the New World Order (2004) (arguing that relying on diplomacy and law rather than force in responding to such threats as posed by political Islam and nonproliferation is being led by European statesmen).

[14] For one critical look, among many, *see* Chalmers A. Johnson, Nemesis: The Last Days of the American Republic (2006).

Africa, and Asia, will spread sufficiently to itself constitute a post-Westphalian alternative form of world order beyond Europe is now quite doubtful. Even so, the regionalization of the world is a possibility worthy of attention even if only to illuminate "the Grotian moment" is generating rival responses designed to provide the world with a post-Westphalian form of global governance. Implicit here is the idea that the state-centric world order that evolved out of the Westphalian peace settlement was a form of global governance that generally seemed successful until the outbreak of the world wars of the prior century, as dramatized by the development and use of atomic bombs in 1945. Despite a certain success from the perspective of dominant elites, there was always much to lament about Westphalian global governance, including providing sanctuaries for "human wrongs" under the rubric of sovereign rights and more or less legitimating both the war system and colonialism.[15]

There is little doubt that the combination of opportunity and danger created by the end of the Cold War and the collapse of the Soviet Union encouraged the neoconservative imaginary to formulate a grand strategy based on global dominance.[16] The 2000 election of George W. Bush as president and the 9/11 attacks enabled this neoconservative blueprint for grand strategy to morph into a political project that became the centerpiece of the "war on terror."[17] This ideological set of moves can be considered from the perspective of global governance as a means to overcome the anarchic character of world order given the globalization and transnationalization of security. It is within this historical and ideological setting that the neoconservative leadership of the United States has tried to solve the crisis of global governance by opting for the "empire" model of world order.[18] The form of empire pursued was definitely distinctive and unlike all historical empires in important re-

[15] For a devastating critique along these lines *see* Ken Booth, *Human Wrongs in International Society,* 71 JOURNAL OF INT'L AFFAIRS 10326 (1995).

[16] For a perspective prior to the Bush presidency and 9/11, *see* Thomas Donnelly, Project for the New American Century, Rebuilding America's Defenses: Strategy, Forces and Resources for a New Century (2000), *available at* http://www.newamericancentury.org/ RebuildingAmericasDefenses.pdf; for a sense of the neoconservative tenor of the Bush presidency, *see* James Mann, Rise of the Vulcans: The History of Bush's War Cabinet (2004).

[17] Nat'l Sec. Council, The National Security Strategy of the United States of America (2002 & rev. ed. 2006), *available at* http://www.whitehouse.gov/nsc /nss.pdf; for a more popular presentation of the same approach, see David Frum & Richard Perle, An End to Evil: How to Win the War on Terror (2003).

[18] There is much recent literature on all sides of this issue. Among the important contributions are: Andrew J. Bacevich, American Empire: The Realities and Consequences of U.S. Diplomacy (2002); Niall Ferguson, Colossus: The Price of America's Empire (2004); Michael Hardt & Antonio Negri, Empire (2001); and Rashid Khalidi, Resurrecting Empire: Western Footprints and America's Perilous Path in the Middle East (2005).

spects. This American way of empire combined a rhetoric of respect for the political independence and territorial integrity of foreign states with a set of security claims of global dimensions that refused to acknowledge any boundaries on its authority and capacity to use force. It also gave unprecedented emphasis to a call for democratic constitutionalism at the state level, even selectively justifying intervention and regime change to rid countries of dictatorial rule. It resorted to aggressive war and exercised extraterritorial authority to implement its counterterrorist foreign policy. Aside from its militarism, it might be difficult to disentangle neoconservative visionary geopolitics as it has been enacted during the Bush presidency from other less provocative ways of establishing American control of world politics in a manner that was also arguably of an imperial character.[19] Imperial geopolitics are perhaps most clearly expressed by the relationship of the United States Government to international law and to the United Nations. International law and the UN due to their potentiality as well as their reality are anti-imperial, clarifying thereby crucial aspects of what, in contrast to empire, a global democracy would entail. Global democracy would certainly entail some kind of respected institutional presence that effectively provides alternatives to war in addressing international disputes, particularly with regard to those issues that touch on vital interests of governments and their citizens. Global democracy would also engender a political culture of respect for the kinds of restraints on the behavior of those states that arise from long diplomatic experience and are then encoded in agreements among governments and other international actors to establish obligatory standards of behavior. As such, it would override the insistence of American leaders on unilateral prerogatives with respect to the use of force, so vividly expressed by President Bush in the 2004 State of the Union Address. In sum, he stated that the United States will never ask for a permission slip whenever its security is at stake.[20] The intention as stated, which was greeted by thunderous bipartisan applause, amounted to a crude insistence that this country, and only this country, retained the discretion to wage war without reference to either the authority of the United Nations or the constraints of international law. This is expressive of a unilateralism that is the decisive repudiation, or the decisive sign of a repudiation, of a commitment to a law governed way of addressing international political behavior.

A repudiation of such unilateralism does not mean a commitment to a legalistic view of the role of international law in our present world. One can appreciate that there may be occasions where the tension between the surviv-

---

[19] On the case for continuity, *see* Neil Smith, The Endgame of Geopolitics (2005).

[20] President George W. Bush, 2004 State of the Union Address (Jan. 20, 2004), *available at* http://www.whitehouse.gov/stateoftheunion/2004/index.html.

al and security of the state and the general prior understanding of international law appear to be in conflict and pose difficult moral, legal, and political choices for national leaders. Recognizing such a possibility of deviating from strict legal strictures still contrasts with the imperial mode that in principle, rather than under existential pressures, repudiates the very idea of constraints on warmaking derived from standards and procedures external to the sovereign state.

As important as is adherence to the Rule of Law with respect to war and peace issues for the establishment of humane forms of global governance, it is not at all synonymous with what we mean when we talk about global democracy. It is my intention to try to provide some introductory understanding of what it is that global democracy would entail, in terms of the organization of the world. In his pioneering work on "cosmopolitan democracy," Daniele Archibugi has argued persuasively that global democracy cannot be properly apprehended as the extension of democracy as it has functioned on the level of the territorial sovereign state to the global level.[21] If global democracy is guided by statist experience, the logical culmination of advocacy of global democracy would be support for a world state and a world government. It is important to understand that this kind of global statism is one possible way of actualizing a commitment to global democracy, but it is probably not the most plausible way and it is certainly, from the perspective of the present, not the most desirable way. It would pose great dangers of world tyranny and world anarchy that would be highly unlikely to produce a form of global governance that could be called "humane." Also, transition to world government seems politically infeasible to such an extent that its endorsement is quickly dismissed as "utopian," that is, unattainable. Although we cannot peer into the future to discern what pathways to global governance will open up under a variety of circumstances, it does not seem useful to give serious attention to world government, whether proceeding from perspectives of global governance or global democracy.[22]

Accordingly, I would like to discuss in a preliminary way some of the developments during the last two decades that seem to be groping toward a set of political outcomes that could culminate over time in a type of global governance that it would be reasonable at some point to call global democracy. We remain very far removed from such a goal at the present time, but this

---

[21] *See generally* COSMOPOLITAN DEMOCRACY: AN AGENDA FOR A NEW WORLD (Daniele Archibugi & David Held eds., 1995); more comprehensively, DEBATING COSMOPOLITICS (Daniele Archibugi et al. eds., 2003).

[22] For the last comprehensive, ambitious proposal for world government, *see* GRENVILLE CLARK & LOUIS B. SOHN, WORLD PEACE THROUGH WORLD LAW: TWO ALTERNATIVE PLANS (3d ed. enlarged 1966) (1958).

should not blind us to a series of important initiatives that point beyond Westphalia without reliance on imperial prerogatives.

The first of these initiatives that deserve mention are the UN Global Conferences that were held particularly in the 1990's. I regard these public events as experiments in global democracy and as the birthing of global civil society.[23] The conferences provided arenas within which nongovernmental organizations, as representatives of civil society, had a number of opportunities. They were able to participate in dialogues that included governments and to develop transnational civil society networks. The strong media presence at these conferences, together with access to the Internet, enabled much greater visibility for civil society perspectives, so much so that this aggregation of influence was sometimes even referred to as being "the second superpower" active in the world after the Cold War.[24] This form of democratic participation by the peoples of the world within global arenas was definitely something new and hopeful. I would argue that it was precisely the success of these experiments that led to a geopolitical backlash that closed off this pathway to global democracy and humane global governance. The major states were not ready to yield their primacy to populist forces expressive of what the peoples of the world demanded and desired.

A second area that I think is extremely relevant and important is the previously mentioned experience of the European Union, also a political experiment intent on moving the theory and practice of democracy beyond the nation-state and establishing a political community that is only indirectly based on state sovereignty. As with global democracy, the EU has paused in its evolution, with its future in doubt. Part of a hopeful scenario for the emergence of global democracy depends on the emergence of democratic forms of regional governance that moderate or even neutralize the turn in the early Twenty-first Century toward global empire.

A third area that points toward global democracy is what I would call "the new internationalism." This kind of post-Westphalian diplomacy was most clearly exhibited in the extraordinary movements during the 1990s resulting in the adoption of an Anti-Personnel Landmines Treaty and the establishment of the International Criminal Court ("ICC"). The defining novel feature of this new internationalism was active and very effective coalitions between clusters of nongovernmental actors and governments of states. This innovative diplomacy was able to overcome the concerted geopolitical objections of the most powerful nations, notably the United States itself, but also China and Russia, to produce new authoritative norms, procedures, and institutions

---

[23] *See generally* MARY KALDOR, GLOBAL CIVIL SOCIETY: AN ANSWER TO WAR (2003).

[24] James F. Moore, *The Second Superpower Rears Its Beautiful Head*, http://cyber.law.harvard.edu/people/jmoore/secondsuperpower.pdf (Mar. 31, 2003).

for international society. Whether the refusal of leading states will doom these efforts remains to be seen. Already, in relation to the ICC, the United States, so determined to oppose, yielded to pressures to encourage the indictment of Sudanese officials alleged to be responsible for crimes against humanity in the context of Darfur. As with the UN global conferences, this kind of new internationalism establishes a mode of democratic participation for the peoples of the world, independent of governmental representation in shaping the realities of global governance.

A fourth initiative involves the activation of national judicial bodies to implement international legal standards. In the context of criminal accountability, this initiative is described beneath the rubric of "universal jurisdiction." This initiative is perhaps best illustrated by the Pinochet litigation that commenced during 1998 in Britain. The Chilean dictator was indicted by a Spanish court, later detained in Britain where extradition hearings were held, and convicted in a historic judgment rendered by the highest British court, the Law Lords.[25] The importance here is that the weakness of the global institutional structure is complemented by a more active judicial role that gives substance to international standards by relying on national judicial institutions to implement universal legal norms. In other words, if national courts become enforcement agencies for international norms, particularly with respect to holding leaders of sovereign states responsible for the crimes against humanity and other crimes of state, there emerges a sense of global governance guided by a set of minimum constraints on the highest officials governing sovereign states.

Again, the challenge to Westphalian modes of geopolitics has provoked a backlash. Belgian laws that were the most revolutionary with respect to universal jurisdiction led to such a strong hostile reaction by the U.S. government accompanied by threats to move NATO headquarters and take other steps. Belgium relented by amending its laws, substantially renouncing its earlier embrace of universal jurisdiction. But all is far from lost. Leading political figures, including Henry Kissinger, have reported changed travel plans for fear of being indicted. There is currently pending with a German prosecutor a complaint against Donald Rumsfeld for his role in the practice of torture at Abu Ghraib. It is likely that geopolitics will prevail, and that the German court will ignore its own law and the strong evidence, but there is a growing sense that global governance depends on establishing the accounta-

---

[25] *See* Richard A. Falk, *Assessing the Pinochet Litigation: Whither Universal Jurisdiction?*, *in* UNIVERSAL JURISDICTION: NATIONAL COURTS AND THE PROSECUTION OF SERIOUS CRIMES UNDER INTERNATIONAL LAW 97 (Stephen Macedo ed., 2004); *see also* NAOMI ROHT-ARRIAZA, THE PINOCHET EFFECT: TRANSNATIONAL JUSTICE IN THE AGE OF HUMAN RIGHTS (2005).

bility of leaders with respect to international criminal law. Those who act on behalf of powerful countries accept such accountability in relation to their adversaries, such as Slobodan Milosevic and Saddam Hussein, but not with respect to themselves.

A fifth initiative has been championed by Andrew Strauss and me, namely the proposal, in its various forms, to establish a global peoples assembly.[26] Symbolically and substantively this initiative recognizes the crucial importance of people participating in a direct manner in the institutional operations of global governance. The initiative presupposes that governmental representation of people, as in the United Nations and global diplomacy, is insufficient. This democratizing demand has proved controversial, but has become over time accepted and successful in the European setting. The European Parliament has finally established itself and been acknowledged as an integral operating part of the European Union and a fundamental element in moves toward European democracy. Much more could be said about the importance and feasibility of a global peoples' parliament. As an undertaking, the project to establish an international criminal court seems now far less utopian than it did in the early 1990's.

A sixth initiative is the existence of tribunals formed by civil society itself. The World Tribunal on Iraq ("WTI") that was held in Istanbul in June 2005 was a very powerful and comprehensive assessment of the status under international law of the American invasion and occupation of Iraq. The WTI included fifty-four presentations to a jury of conscience that drew on the expert knowledge of prominent international lawyers and international political experts as well as received emotionally powerful testimony from notable Iraqis. The primary justification for the creation of such a tribunal was to fill the gap created by the unwillingness and inability of either governments in international society or the United Nations to act meaningfully on behalf of fundamental norms of international law.

The WTI was impressive for a number of reasons. First, it was the culmination of twenty earlier civil society tribunals held all over the world on the Iraq War. Second, the WTI represented the first time that civil society was mobilized on a global basis to oppose a war that was so widely perceived throughout the world as illegal and an example of aggressive war of the sort prohibited by the UN charter. Third, the WTI exhibited an entirely new phenomenon that might be called "moral globalization," a spontaneous expression of support for the implementation of agreed fundamental norms, the constitutional basis of humane global governance, and a corresponding repu-

---

[26] Richard Falk & Andrew Strauss, *On the Creation of a Global Peoples Assembly: Legitimacy and the Power of Popular Sovereignty*, 36 STANFORD L. REV. 191220 (2000); Falk & Strauss, *Toward Global Parliament*, 80 FOREIGN AFFAIRS 80: 212 (2001).

diation of geopolitical claims of entitlement with respect to war as a political option.

The last initiative I will mention is the dependence of a movement toward global democracy upon the education of citizens, especially here in the United States. More generally, it is a vital component of the educational responsibility of institutions of higher learning throughout the world to prepare young people for engaged citizenship in this young Twenty-first Century. Furthermore, I believe the prospect of achieving global democracy depends on internalizing the sort of values and global outlook that would allow that kind of political development beyond the sovereign state to take place. I think that two areas of educational emphasis would be particularly valuable at this stage of history. One is the importance of making citizens of this country and of other countries much more familiar with the relevance of a culture of human rights as part of their own development as members of any political community entitled to all aspects of human dignity. It seems clear that to the extent that human rights are internalized as part of legitimate governance at any level of societal organization, it will facilitate a popular acceptance of the need for the construction of global democracy by consensual means.

The second educational priority is currently more controversial, but at least as necessary. It involves making a pedagogy of peace and human security an important part of the learning experience of every young person.[27] In my view, available evidence suggests the increasing dysfunctionality of war as an instrument for the resolution of conflict. On this basis, it is a virtual imperative to explore alternatives to war and political violence. Our educational experience should challenge the political and moral imagination of students by considering the benefits of reliance on nonviolent politics as the foundation of global security, reform, and justice in the world. The essence of global democracy involves a shift in expectations from a geopolitics of force to a geopolitics of dialogue and persuasion.

The goals of global democracy and humane global governance certainly seem remote from current patterns of behavior in all sectors of the world. The position taken here is that without such normative horizons, we will be enveloped by the storm clouds now gathering so menacingly as to defy disbelief. Hope begins when we have the moral courage to transcend what seems possible by what seems necessary and desirable. I think the changing parameters of debate on climate change, facing that "inconvenient truth," is an encouraging sign of an emerging receptivity to an acceptance of constraints on behavior for the sake of a humane future.

On the Authors

Richard Falk

Richard Falk is the Albert G. Milbank Professor Emeritus of International Law at Princeton University and Visiting Distinguished Professor in Global and International Studies at the University of California, Santa Barbara. He received his B.S. from the Wharton School, University of Pennsylvania, his L.L.B. from Yale Law School, and his J.S.D. from Harvard University.

During his career Professor Falk has written hundreds of articles and book chapters and has authored or coauthored over 50 books, including most recently *Achieving Human Rights*. Other notable works include: *Religion and Humane Global Governance*; *Human Rights Horizons*; *On Humane Governance: Toward a New Global Politics*; *Explorations at the Edge of Time*; *Revolutionaries and Functionaries*; *The Promise of World Order*; *Indefensible Weapons*; *Human Rights and State Sovereignty*; *A Study of Future Worlds*; *This Endangered Planet*; and he was the co-editor of *Crimes of War*.

In 2001 he served on a three person Human Rights Inquiry Commission for the Palestine Territories that was appointed by the United Nations, and in 2000 he served on the Independent International Commission on Kosovo. In 2008, he was appointed by the U.N. Human Rights Council to be the special *Rapporteur* for the Occupied Palestinian Territories.

Professor Falk serves as Chair of the Nuclear Age Peace Foundation's Board of Directors and as honorary Vice President of the American Society of International Law. He is also a member of the Editorial Board of *The Nation*. Professor Falk also acted as counsel to Ethiopia and Liberia in the Southwest Africa Case before the International Court of Justice. He was nominated for the Nobel Peace Prize in 2008, 2009 and 2010.

Andrew Strauss

Andrew Strauss is a Distinguished Professor of Law at Widener University School of Law. He earned his Bachelor of Arts from Princeton University's Woodrow Wilson School of Public and International Affairs and his Juris Doctorate from New York University School of Law.

Professor Strauss is co-author (with Burns Weston, Richard Falk and Hilary Charlesworth) of the fourth edition of *International Law and World Order*, a leading international law textbook. His articles have appeared in international journals such as *Foreign Affairs, The Harvard Journal of International Law*, and *The Stanford Journal of International Law*. He is also a frequent public commentator on matters of international law and policy with articles appearing in such publications as *The International Herald Tribune, The Nation*, and *The Financial Times*. Among his contributions to the broadcast media, his radio commentaries have been aired on Public Radio International's *Marketplace*.

Professor Strauss was a Fulbright Scholar in Ecuador where he studied tribal politics in the Amazon. He has been a Visiting Professor at Notre Dame Law School, a lecturer at the European Peace University in Schlaining Austria, and he has taught Singaporean constitutional law on the law faculty of the National University of Singapore. He has served as the Director of the Geneva/Lausanne International Law Institute and the Nairobi International Law Institute. He has also been an Honorary Fellow at New York University School of Law's Center for International Studies.

Professor Strauss is internationally active in many civic and professional organizations. He has conducted human rights investigative missions and been a consultant to both Human Rights Watch and Human Rights First. Professor Strauss is a member of the Consultants Working Group of the Climate Legacy Initiative. He is a member of the International Advisors Group of the One World Trust, is on the Advisory Council of the Center for U.N. Reform Education, and is a fellow of the Citizens for Global Solutions World Federalist Institute.

Index